Designing Sociable Robots

Designing Sociable Robots

Cynthia L. Breazeal

A Bradford Book
The MIT Press
Cambridge, Massachusetts
London, England

First MIT Press paperback edition, 2004

This book was set in Times Roman by the author and Interactive Composition Corporation using LaTeX.

Printed and bound in the United States of America.

Library of Congress Cataloging-in-Publication Data

Breazeal, Cynthia L.
    Designing sociable robots / Cynthia L. Breazeal.
    p.   cm.—(Intelligent robots and autonomous agents)
    "A Bradford book."
    ISBN 978-0-262-02510-2 (hc. : alk. paper)—978-0-262-52431-5 (pb.)
    1. Human-machine systems. 2. Artificial intelligence. I. Title. II. Series.

TA167.B74 2001
006.3—dc21   2001034233

To our children of the future, organic or synthetic

The twenty-first century may very well be the century of the robot.
—Katherine Hayles, from the documentary *Into the Body*

Contents

Preface

I remember seeing the movie *Star Wars* as a little girl. I remember being absolutely captivated and fascinated by the two droids, R2-D2 and C-3P0. Their personalities and their antics made them compelling characters, far different from typical sci-fi robots. I actually cared about these droids, unlike the computer HAL, from Arthur C. Clarke's book *2001: A Space Odyssey*, whose cool intelligence left me with an eerie feeling. I remember the heated debates among my classmates about whether the droids were real or not. Some would argue that because you could see the wires in C-3P0's torso that it must be a real robot. Alas, however, the truth was known. They were not real at all. They existed only in the movies. I figured that I would never see anything like those robots in my lifetime.

Many years later I found myself at the MIT Artificial Intelligence Lab with the opportunity to work with Professor Rod Brooks. He told me of autonomous robots, of their biological inspiration, all very insect-like in nature. I remember thinking to myself that this was it—these kinds of robots were the real-life precursors to the *Star Wars* droids of my childhood. I knew that this was the place for me. Trained in engineering and the sciences, I began to specialize in robotics and artificial intelligence. While working at the MIT Artificial Intelligence Lab, my colleagues and I have created a wide assortment of autonomous robots, ranging from insect-like planetary micro-rovers to upper-torso humanoids, their behavior mirroring that of biological creatures. I developed a deep appreciation for the insights that science as well as art have to offer in building "living, breathing" robots. As a well-seasoned researcher, I began to build a robot in the image of my childhood dream. Its name is Kismet, and it is largely the subject of this book.

Beyond the inspiration and implementation of Kismet, this book also tries to define a vision for sociable robots of the future. Taking R2-D2 and C-3P0 as representative instances, a sociable robot is able to communicate and interact with us, understand and even relate to us, in a personal way. It is a robot that is socially intelligent in a human-like way. We interact with it as if it were a person, and ultimately as a friend. This is the dream of a sociable robot. The field is in its infancy, and so is Kismet.

The year 2001 has arrived. The vast majority of modern robots are sophisticated tools, not synthetic creatures. They are used to manufacture cars more efficiently and quickly, to explore the depths of the ocean, or to exceed our human limitations to perform delicate surgery. These and many other applications are driven by the desire to increase efficiency, productivity, and effectiveness in utilitarian terms, or to perform tasks in environments too hazardous for humans. They are valued for their ability to carry out tasks without interacting with people.

Recently, robotic technologies are making their way into society at large, commercialized as toys, cyber-pets, or other entertainment products. The development of robots for domestic and healthcare purposes is already under way in corporate and university research labs. For these applications, the ability to interact with a wide variety of people in a natural, intuitive, and enjoyable manner is important and valuable. We are entering a time when socially savvy robots could achieve commercial success, potentially transforming society.

But will people interact socially with these robots? Indeed, this appears to be the case. In the field of human computer interaction (HCI), experiments have revealed that people unconsciously treat socially interactive technologies like people, demonstrating politeness, showing concern for their "feelings," etc. To understand why, consider the profound impact that overcoming social challenges has had on the evolution of the human brain. In essence, we have evolved to be *experts* in social interaction. Our brains have changed very little from that of our long-past ancestors, yet we must deal with modern technology. As a result, if a technology behaves in a socially competent manner, we evoke our evolved social machinery to interact with it. Humanoid robots are a particularly intriguing technology for interacting with people, given the robots' ability to support familiar social cues.

Hence, it makes practical sense to design robots that interact with us in a familiar way. Humanizing the interface and our relationship with robots, however, depends on our conceptions of human nature and what constitutes human-style social interaction. Accordingly, we must consider the specific ways we understand and interact with the social world. If done well, these robots will support our social characteristics, and our interactions with them will be natural and intuitive. Thus, in an effort to make sociable robots familiar to people, they will have to be socially intelligent in a human-like way.

There are a myriad of reasons—scientific, philosophical, as well as practical—for why social intelligence is important for robots that interact with people. Social factors profoundly shaped our evolution as a species. They play a critical role in our cognitive development, how we learn from others, how we communicate and interact, our culture, and our daily lives as members of society. For robots to be a part of our daily lives, they must be responsive to us and be able to adapt in a manner that is natural and intuitive for us, not vice versa. In this way, building sociable robots is also a means for understanding human social intelligence as well—by providing testbeds for theories and models that underlie our social abilities, through building engaging and intelligent robots that assist in our daily lives as well as learn from us and teach us, and by challenging us to reflect upon the nature of humanity and society. Robots should not supplant our need to interact with each other, but rather should support us in our quest to better understand ourselves so that we might appreciate, enhance, and celebrate our humanity and our social lives.

As the sociality of these robots begins to rival our own, will we accept them into the human community? How will we treat them as they grow to understand us, relate to us, empathize with us, befriend us, and share our lives? Science fiction has long challenged us to ponder these questions. Vintage science fiction often portrays robots as sophisticated appliances that people command to do their bidding. *Star Wars,* however, endows mechanical droids with human characteristics. They have interesting personalities. They fear personal harm but will risk their lives to save their friends. They are not appliances, but servants, arguably even slaves that are bought and sold into servitude. The same holds true for the androids of Philip K. Dick's *Do Androids Dream of Electric Sheep?,* although their appearance and behavior are virtually indistinguishable from their human counterparts. The android, Data, of the television series *Star Trek: The Next Generation* provides a third example of an individualized robot, but with an unusual social status. Data has a human-like appearance but possesses super-human strength and intellect. As an officer on a starship, Data outranks many of the humans onboard. Yet this android's personal quest is to become human, and an essential part of this is to develop human-like emotions.

It is no wonder that science fiction loves robots, androids, and cyborgs. These stories force us to reflect upon the nature of being human and to question our society. Robots will become more socially intelligent and by doing so will become more like us. Meanwhile we strive to enhance ourselves by integrating technology into our lives and even into our bodies. Technological visionaries argue that we are well on the path to becoming cyborgs, replacing more and more of our biological matter with technologically enhanced counterparts. Will we still be human? What does it mean to be a person? The quest of building socially intelligent robots forces us to examine these questions even today.

I've written this book as a step on the way to the creation of sociable robots. A significant contribution of the book is the presentation of a concrete instance of a nascent sociable robot, namely Kismet.

Kismet is special and unique. Not only because of what it can do, but also because of how it makes you feel. Kismet connects to people on a physical level, on a social level, and on an emotional level. It is jarring for people to play with Kismet and then see it turned off, suddenly becoming an inanimate object. For this reason, I do not see Kismet as being a purely scientific or engineering endeavor. It is an artistic endeavor as well. It is my masterpiece. Unfortunately, I do not think anyone can get a full appreciation of what Kismet is merely by reading this book. To aid in this, I have included a CD-ROM so that you can see Kismet in action.* Yet, to understand the connection this robot makes with so many people, I think you have to experience it first hand.

---

* *In this printing, the CD-ROM has been omitted in favor of online access to its files. For all references to the CD-ROM, please use the online resources available at https://mitpress.mit.edu/books/designing-sociable-robots.*

Acknowledgments

The word "kismet" is Turkish, meaning "destiny" or "fate." Ironically, perhaps, I was destined to build a robot like Kismet. I could have never built Kismet alone, however. Throughout this book, I use the personal pronoun "I" for easier reading, but in reality, this project relied on the talents of many, many others who have contributed ideas, shaped my thoughts, and hacked code for the little one. There are so many people to give my heartfelt thanks. Kismet would not be what it is today without you.

First, I should thank Prof. Rod Brooks. He has believed in me, supported me, and given me the freedom to pursue not one but several ridiculously ambitious projects. He has been my mentor and my friend. As a robotics visionary he has always encouraged me to dream large and to think out of the box. I honestly cannot think of another place in the world, working for anyone else, where I would have been given the opportunity to even attempt what I have accomplished in this lab. Of course, opportunity often requires resources and money, so I want to gratefully acknowledge those who funded Kismet. Support for Kismet was provided in part by an ONR Vision MURI Grant (No. N00014-95-1-0600), and in part by DARPA/ITO under contract DABT 63-99-1-0012. I hope they are happy with the results.

Next, there are those who have put so much of their time and effort into making Kismet tick. There are my colleagues in the Humanoid Robotics Group at the MIT Artificial Intelligence Lab who have worked with me to give Kismet the ability to see and hear. In particular, I am indebted to Brian Scassellati, Paul Fitzpatrick, Lijin Aryananda, and Paulina Varchavskaia. Kismet wouldn't hear a thing if it were not for the help of Jim Glass and Lee Hetherington of the Spoken Language Systems Group in the Laboratory for Computer Science at MIT. They were very generous with their time and support in porting the SLS speech recognition code to Kismet. Ulysses Gilchrist improved upon the mechanical design of Kismet, adding several new degrees of freedom. I would like to acknowledge Jim Alser at Tech Optics for figuring out how to make Kismet's captivating blue eyes. It would not the same robot without them.

I've had so many useful discussions with my colleagues over the years in the Mobile Robots Group and the Humanoid Robotics Group at MIT. I've picked Juan Velasquez's brain on many occasions about theories on emotion. I've cornered Robert Irie again and again about auditory processing. I've bugged Matto Marjanovic throughout the years to figure out how to build random electronic stuff. Kerstin Dautenhahn and Brian Scassellati are kindred spirits with the shared dream of building socially intelligent robots, and our discussions have had a profound impact on the ideas in this book.

Bruce Blumberg was the one who first opened my eyes to the world of animation and synthetic characters. The concepts of believability, expressiveness, and audience perception are so critical for building sociable machines. I now see many strong parallels between his field and my own, and I have learned so much from him. I've had great discussions with Chris Kline and Mike Hlavac from Bruce's Synthetic Characters Group at the MIT Media

Lab. Roz Picard provided many useful comments and suggestions for the design of Kismet's emotion system and its ability to express its affective state through face and voice. Justine Cassell's insights and knowledge of face-to-face communication have had significant impact on the design of Kismet's communication skills. Discussions with Anne Foerst and Brian Knappenberger have helped me to contemplate and appreciate the questions of personhood and human identity that are raised by this work. I am grateful to Sherry Turkle for being such a strong supporter of my research over the years. Discussions with Sandy Pentland have encouraged me to explore new paradigms for socially intelligent robots, beyond creatures to include intelligent, physically animated spaces and wearable robots. I am indebted to those who have read numerous drafts of this work including Roz Picard, Justine Cassell, Cory Kidd, Rod Brooks, and Paul Fitzpatrick. I have tried to incorporate their many useful comments and suggestions.

I also want to thank Robert Prior at The MIT Press for his enthusiastic support of this book project. Abigail Mieko Vargus copyedited numerous versions of the book draft and Mel Goldsipe of The MIT Press was a tremendous help in making this a polished final product. I want to extend my grateful acknowledgement to Peter Menzel[1] and Sam Ogden[2] for graciously allowing me to use their beautiful images of Kismet and other robots. MIT Video Productions did a beautiful job in producing the accompanying CD-ROM[3] using a significant amount of footage courtesy of The MIT Museum.

Finally, my family and dear friends have encouraged me and supported me through my personal journey. I would not be who I am today without you in my life. My mother Juliette, my father Norman, and my brother William have all stood by me through the best of times and the worst of times. Their unconditional love and support have helped me through some very difficult times. I do my best to give them reason to be proud. Brian Anthony has encouraged me when I needed it most. He often reminded me, "Life is a process—enjoy the process." There have been so many friends, past and present. I thank you all for sharing yourselves with me and I am deeply grateful.

1. Images ©2000 Peter Menzel from the book *Robo sapiens: Evolution of a New Species* by Peter Menzel and Faith D'Alusio, a Material World Book published by The MIT Press. Fall 2000.

2. Images © Sam Ogden.

3. *Designing Sociable Robots* CD-ROM ©2002 Massachusetts Institute of Technology.

Sources

This book is based on research that was previously reported in the following publications.

B. Adams, C. Breazeal, R. Brooks, P. Fitzpatrick, and B. Scassellati, "Humanoid Robots: A New Kind of Tool," in *IEEE Intelligent Systems, Special Issue on Humanoid Robotics,* 15:4, 25–31, (2000).

R. Brooks, C. Breazeal (Ferrell), R. Irie, C. Kemp, M. Marjanovic, B. Scassellati, M. Williamson, "Alternative essences of intelligence," in *Proceedings of the Fifteenth National Conference on Artificial Intelligence (AAAI98).* Madison, WI, 961–967, (1998).

C. Breazeal, "Affective interaction between humans and robots," in *Proceedings of the 2001 European Conference on Artificial Life (ECAL2001).* Prague, Czech Rep., 582–591, (2001).

C. Breazeal, "Believability and readability of robot faces," in *Proceedings of the Eighth International Symposium on Intelligent Robotic Systems (SIRS2000).* Reading, UK, 247–256, (2000).

C. Breazeal, "Designing Sociable Robots: Issues and Lessons," in K. Dautenhahn, A. Bond, and L. Canamero (eds.), *Socially Intelligent Agents: Creating Relationships with Computers and Robots,* Kluwer Academic Press, (in press).

C. Breazeal, "Emotive qualities in robot speech," in *Proceedings of the 2001 International Conference on Intelligent Robotics and Systems (IROS2001).* Maui, HI, (2001). [CD-ROM proceedings.]

C. Breazeal, "A motivational system for regulating human-robot interaction," in *Proceedings of the Fifteenth National Conference on Artificial Intelligence (AAAI98).* Madison, WI, 54–61, (1998).

C. Breazeal, "Proto-conversations with an anthropomorphic robot," in *Proceedings of the Ninth IEEE International Workshop on Robot and Human Interactive Communication (Ro-Man2000).* Osaka, Japan, 328–333, (2000).

C. Breazeal, *Sociable Machines: Expressive Social Exchange between Humans and Robots,* Ph.D. thesis, Massachusetts Institute of Technology, Department of Electrical Engineering and Computer Science, Cambridge, MA, (2000).

C. Breazeal and L. Aryananda, "Recognizing affective intent in robot directed speech," in *Autonomous Robots,* 12:1, 83–104, (2002).

C. Breazeal, A. Edsinger, P. Fitzpatrick, B. Scassellati, and P. Varchavskaia, "Social Constraints on Animate Vision," in *IEEE Intelligent Systems, Special Issue on Humanoid Robotics,* 15:4, 32–37, (2000).

C. Breazeal, P. Fitzpatrick, and B. Scassellati, "Active vision systems for sociable robots," in K. Dautenhahn (ed.), *IEEE Transactions on Systems, Man, and Cybernetics,* 31:5, (2001).

C. Breazeal and A. Foerst, "Schmoozing with robots, exploring the boundary of the original wireless network," in *Proceedings of the 1999 Conference on Cognitive Technology (CT99)*. San Francisco, CA, 375–390, (1999).

C. Breazeal and B. Scassellati, "Challenges in Building Robots That Imitate People," in K. Dautenhahn and C. Nehaniv (eds.), *Imitation in Animals and Artifacts*, MIT Press, (in press).

C. Breazeal and B. Scassellati, "A context-dependent attention system for a social robot," in *Proceedings of the Sixteenth International Joint Conference on Artificial Intelligence (IJCAI99)*. Stockholm, Sweden, 1146–1151, (1999).

C. Breazeal and B. Scassellati, "How to build robots that make friends and influence people," in *Proceedings of the 1999 IEEE/RSJ International Conference on Intelligent Robots and Systems (IROS99)*. Kyonjiu, Korea, 858–863, (1999).

C. Breazeal (Ferrell) and B. Scassellati, "Infant-like social interactions between a robot and a human caregiver," in K. Dautenhahn (ed.), *Adaptive Behavior*, 8:1, 47–72, (2000).

1 The Vision of Sociable Robots

What is a sociable robot? It is a difficult concept to define, but science fiction offers many examples. There are the mechanical droids R2-D2 and C-3PO from the movie *Star Wars* and the android Lt. Commander Data from the television series *Star Trek: The Next Generation*. Many wonderful examples exist in the short stories of Isaac Asimov and Brian Aldiss, such as the robots Robbie (Asimov, 1986) and David (Aldiss, 2001). For me, a sociable robot is able to communicate and interact with us, understand and even relate to us, in a personal way. It should be able to understand us and itself in social terms. We, in turn, should be able to understand it in the same social terms—to be able to relate to it and to empathize with it. Such a robot must be able to adapt and learn throughout its lifetime, incorporating shared experiences with other individuals into its understanding of self, of others, and of the relationships they share. In short, a sociable robot is socially intelligent in a human-like way, and interacting with it is like interacting with another person. At the pinnacle of achievement, they could befriend us, as we could them. Science fiction illustrates how these technologies could enhance our lives and benefit society, but it also warns us that this dream must be approached responsibly and ethically, as portrayed in Philip K. Dick's *Do Androids Dream of Electric Sheep* (Dick, 1990) (made into the movie *Blade Runner*).

1.1 Why Sociable Robots?

Socially intelligent robots are not only interesting for science fiction. There are scientific and practical reasons for building robots that can interact with people in a human-centered manner. From a scientific perspective, we could learn a lot about ourselves from the process of building socially intelligent robots. Our evolution, our development from infancy to adulthood, our culture from generation to generation, and our day-to-day existence in society are all profoundly shaped by social factors (Vygotsky et al., 1980; Forgas, 2000; Brothers, 1997; Mead, 1934). Understanding our sociality is critical to understanding our humanity.

Toward this goal, robots could be used as experimental testbeds for scientific inquiry (Adams et al., 2000). Computational models of our social abilities could be implemented, tested, and analyzed on robots as they participate in controlled social scenarios. In this way, robots could potentially be used in the same studies and experiments that scientists use to understand human social behavior. Robot data could be compared with human performance under similar conditions. Differences between the two could be used to refine the models and inspire new experiments. Furthermore, given a thorough understanding of the implementation, parameters of the model could be systematically varied to understand their effects on social behavior. By doing so, social behavior disorders could be better understood, which in turn could aid in the development of effective treatments. For instance, autism is regarded as an impairment in the ability to interact with and understand others in social terms. A few efforts are under way to use robots in treatment of autistic children (Dautenhahn, 2000) and to try to understand this impairment by modeling it on robots (Scassellati, 2000b).

As humans, we not only strive to understand ourselves, but we also turn to technology to enhance the quality of our lives. From an engineering perspective, we try to make these technologies natural and intuitive to use and to interact with. As our technologies become more intelligent and more complex, we still want to interact with them in a familiar way. We tend to anthropomorphize our computers, our cars, and other gadgets for this reason, and their interfaces resemble how we interact with each other more and more (Mithen, 1996). Perhaps this is not surprising given that our brains have evolved for us to be experts in social interaction (Barton & Dunbar, 1997).

Traditionally, autonomous robots have been targeted for applications requiring very little (if any) interaction with humans, such as sweeping minefields, inspecting oil wells, or exploring other planets. Other applications such as delivering hospital meals, mowing lawns, or vacuuming floors bring autonomous robots into environments shared with people, but human-robot interaction in these tasks is still minimal. Examples of these robots are shown in figure 1.1.

New commercial applications are emerging where the ability to interact with people in a compelling and enjoyable manner is an important part of the robot's functionality. A couple of examples are shown in figure 1.2. A new generation of robotic toys have emerged, such as Furby, a small fanciful creature whose behavior changes the more children play with it. Dolls and "cyber-pets" are beginning to incorporate robotic technologies as well. For

Figure 1.1
Some examples of applications motivating autonomous robots. To the left is NASA's *Sojourner,* a planetary micro-rover that gathered scientific data on Mars. To the right is a commercial autonomous vacuum-cleaning robot.

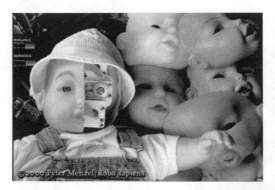

Figure 1.2
Some examples of robots entering the toy and entertainment markets. To the left is iRobot's Bit, a prototype robotic doll that can display a number of facial expressions. To the right is Tiger Electronic's Furby.

Figure 1.3
Some examples of research exploring robots that cooperate with and assist humans. On the left is Sweet Lips, a museum tour guide robot. The right shows NEC's domestic robot prototype.

instance, Hasboro's My Real Baby changes facial expressions according to its "mood," which is influenced by how it is played with. Although the ability of these products to interact with people is limited, they are motivating the development of increasingly life-like and socially sophisticated robots. Someday, these toys might be sophisticated enough to appreciate and foster the social needs and cognitive development of a child.

Companies and universities are exploring new applications areas for robots that assist people in a number of ways (see figure 1.3). For instance, robotic tour guides have appeared in a few museums and are very popular with children (Burgard et al., 1998; Thrun et al., 1999). Honda has developed an adult-sized humanoid robot called P3 and a child-sized version called Asimo. The company is exploring entertainment applications, such as robotic soccer players.[1] Eventually, however, it will be plausible for companies to pursue domestic uses for robots, humanoid or otherwise. For example, NEC is developing a household robot resembling R2-D2 that can help people interact with electronic devices around the house (e.g., TV, computer, answering service, etc.). Health-related applications are also being explored, such as the use of robots as nursemaids to help the elderly (Dario & Susani, 1996; see also `www.cs.cmu.edu/~nursebot`). The commercial success of these robots hinges on their ability to be part of a person's daily life. As a result, the robots must be responsive to and interact with people in a natural and intuitive manner.

It is difficult to predict what other applications the future holds for socially intelligent robots. Science fiction has certainly been a source of inspiration for many of the applications being explored today. As a different twist, what if you could "project" yourself into a physical avatar? Unlike telerobotics or telepresence of today, the robotic "host" would have to be socially savvy enough to understand the intention of the human "symbiont." Then, acting in concert with the human, the robot would faithfully carry out the person's wishes while portraying his/her personality. This would enable people to physically interact with faraway people, an exciting prospect for people who are physically isolated, perhaps bedridden for health reasons.

Another possibility is an artifact that you wear or carry with you. An example from science fiction would be the Primer described in Neal Stephenson's *The Diamond Age* (2000). The Primer is an interactive book equipped with sophisticated artificial intelligence. It is socially aware of the little girl who owns it, can identify her specifically, knows her personally, is aware of her education and abilities, and shapes its lessons to foster her continued growth and development into adulthood. As another possibility, the technology could take the form of a small creature, like a gargoyle, that sits on your shoulder and acts as an information assistant for you.[2] Over time, the gargoyle could adapt to you, learn your preferences, retrieve information for you—similar to the tasks that software agents might carry out while sharing your world and supporting natural human-style interaction. These gargoyles could interact with each other as well, serving as social facilitators to bring people with common interests into contact with each other.

1. Robocup is an organized event where researchers build soccer-playing robots to investigate research questions into cooperative behavior, team strategy, and learning (Kitano et al., 1997; Veloso et al., 1997).

2. Rhodes (1997) talks of a rememberance agent, a continuously running proactive memory aid that uses the physical context of a wearable computer to provide notes that might be relevant in that context. This is a similar idea, but now it is a wearable robot instead of a wearable computer.

1.2   The Robot, Kismet

The goal of this book is to pioneer a path toward the creation of sociable robots. Along the way, I've tried to provide a map of this relatively uncharted area so that others might follow. Toward this goal, the remainder of this chapter offers several key components of social intelligence and discusses what these abilities consist of for these machines. Many of these attributes are derived from several distinguishing characteristics of human social intelligence. From this, I construct a framework and define a set of design issues for building socially intelligent robots in the following chapters. Our journey should be a responsible one, well-conceived and well-intentioned. For this reason, this book also raises some of the philosophical and ethical questions regarding how building such technologies shapes our self-understanding, and how these technologies might impact society. This book does not provide answers but instead hopes to foster discussion that will help us to develop these sorts of technologies in responsible ways.

Aspects of this potentially could be applied to the design of socially intelligent software agents. There are significant differences between the physical world of humans and the virtual world of computer agents, however. These differences impact how people perceive and interact with these two different types of technology, and vice versa. Perhaps the most striking difference is the physical and immediately proximate interactions that transpire between humans and robots that share the same social world. Some issues and constraints remain distinct for these different technologies. For this reason, I acknowledge relevant research in the software agents community, but focus my presentation on the efforts in the robotics domain.

Humans are the most socially advanced of all species. As one might imagine, an autonomous humanoid robot that could interpret, respond, and deliver human-style social cues even at the level of a human infant is quite a sophisticated machine. Hence, this book explores the simplest kind of human-style social interaction and learning, that which occurs between a human infant with its caregiver. My primary interest in building this kind of sociable, infant-like robot is to explore the challenges of building a socially intelligent machine that can communicate with and learn from people.

This is a scientific endeavor, an engineering challenge, and an artistic pursuit. Starting in 1997, my colleagues and I at the MIT Artificial Intelligence Lab began to construct such a robot (see figure 1.4). It is called Kismet, and we have implemented a wide variety of infant-level social competencies into it by adapting models and theories from the fields of psychology, cognitive development, and ethology. This book, a revised version of my doctoral dissertation (Breazeal, 2000c), uses the implementation of Kismet as a case study to illustrate how this framework is applied, how these design issues are met, how scientific and artistic insights are incorporated into the design, and how the work is evaluated. It is a very

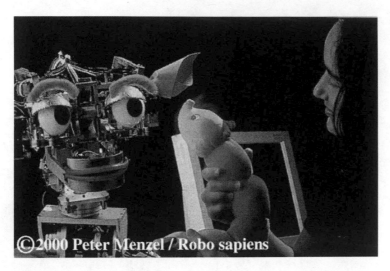

Figure 1.4
Kismet, a sociable "infant" robot being developed at MIT.

ambitious and highly integrated system, running on fifteen networked computers. (If you have not viewed the enclosed CD-ROM, I recommend you do so. I will reference its demos at relevant points as well.) This book reveals the ideas, insights, and inspiration, and technical details underlying Kismet's compelling, life-like behavior. Significant progress has been made, yet much work remains to be done to fully realize the vision of a sociable robot.

1.3   Ingredients of Sociable Robots

As stated in the preface, one goal of building a sociable robot is to gain a scientific under-standing of social intelligence and human sociality. Another goal is to design robots that can interact with people on "human terms." Accordingly, it is important to consider the specific ways in which we understand and interact with the social world. If done well, humans will be able to engage the robot by utilizing their natural social machinery instead of having to overly and artificially adapt their way of interaction. Dautenhahn (1998) identifies a number of characteristics of human social intelligence that should be considered when designing socially intelligent technologies. Much of the discussion in this section (and in the final chapter in section 13.3) is based on the broader issues of human-style social intelligence as presented by Dautenhahn. These key characteristics of human social intelligence have guided my work with Kismet, and the body work presented in this book both instantiates and elaborates upon them.

Being There

Humans are embodied and situated in the social world. We ground our experiences through our body as we interact with the environment and with others. As such, our bodies provide us with a means for relating to the world and for giving our experiences meaning (Lakoff, 1990). Brooks has extensively argued for the importance of embodiment and being situated in the world for understanding and generating intelligent behavior in animals and robots (Brooks, 1990). Socially intelligent robots can better support these human characteristics if they are *embodied* and *socially situated* with people. For this reason, Kismet is a physical robot that interacts with people face-to-face.

Having a body and existing within a shared environment is advantageous for both the robot as well as for those people who interact with it. From the perspective of the robot, its body provides it with a vehicle for experiencing and for interacting with the social world. Further, the robot can interpret these experiences within a social context. From the perspective of a human who interacts with the robot, it is also beneficial for the robot to have a body. Given that humans have evolved to socially interact with embodied creatures, many of our social skills and communication modalities rely on both parties having a body. For instance, people frequently exchange facial expressions, gestures, and shift their gaze direction when communicating with others. Even at a more basic level, people rely on having a point of reference for directing their communication efforts toward the desired individual, and for knowing where to look for communicative feedback from that individual.

The embodiment and situatedness of a robot can take several forms. For instance, the robot could share the same physical space as a person, such as a humanoid robot that communicates using familiar social cues (Brooks et al., 1999). Alternatively, the technology could be a computer-animated agent within a virtual space that interacts with a human in the physical world. Embodied conversational agents (Cassell, 1999a) are a prime example. It is also possible to employ virtual-reality (VR) techniques to immerse the human within the virtual world of the animated agent (Rickel & Johnson, 2000). These robots or animated agents are often humanoid in form to support gestures, facial expressions, and other embodied social cues that are familiar to humans. The nature of the experience for the human varies in each of these different scenarios depending upon the sensing limits of the technologies (such as keyboards, cameras, microphones, etc.); whether the human must be instrumented (e.g., wearing data gloves, VR helmets, etc.); the amount of freedom the person has to move within the space; and the type of display technology employed, be it mechanical, projected on a large screen, or displayed on a computer monitor.

Life-Like Quality

People are attracted to life-like behavior and seem quite willing to anthropomorphize nature and even technological artifacts. We appear biased to perceive and recognize other living beings and are able to do so quite early in our development (Trevarthen, 1979). We tend to interpret behavior (such as self-propelled movement) as being intentional, whether it is demonstrated by a living creature or not (Premack & Premack, 1995). When engaging a non-living agent in a social manner, people show the same tendencies (Reeves & Nass, 1996). Ideally, humans would interact with robots as naturally as they interact with other people. To facilitate this kind of social interaction, robot behavior should reflect life-like qualities. Much attention has been directed to giving Kismet's behavior this quality so that people will engage the robot naturally as a social being.

Living agents such as animals and humans are *autonomous*. They are capable of promoting their survival and performing tasks while negotiating the complexities of daily life. This involves maintaining their desired relationship with the environment, yet they continually change this balance as resources are competed for and consumed. Robots that share a social environment with others must also able to foster their continued existence while performing their tasks as they interact with others in an ever-changing environment.

Autonomy alone is not sufficiently life-like for human-style sociability, however. Interacting with a sociable robot should not be like interacting with an ant or a fish, for instance. Although ants and fish are social species, they do not support the human desire to treat others as distinct personalities and to be treated the same in turn. For this reason, it is important that sociable robots be *believable*.

The concept of believability originated in the arts for classically animated characters (Thomas & Johnston, 1981) and was later introduced to interactive software agents (Bates, 1994). Believable agents project the "illusion of life" and convey personality to the human who interacts with it. To be believable, an observer must be able and willing to apply sophisticated social-cognitive abilities to predict, understand, and explain the character's observable behavior and inferred mental states in familiar social terms. Displaying behaviors such as giving attention, emotional expression, and playful antics enable the human observer to understand and relate to these characters in human terms. Pixar and Walt Disney are masters at creating believable characters, animating and anthropomorphizing nature and inanimate objects from trees to Luxo lamps. An excellent discussion of believability in robots can be found in Dautenhahn (1997, 1998).

Human-Aware

To interact with people in a human-like manner, sociable robots must *perceive* and *understand* the richness and complexity of natural human social behavior. Humans communicate with one another through gaze direction, facial expression, body movement, speech, and language, to name a few. The recipient of these observable signals combines them with knowledge of the sender's personality, culture, past history, the present situational context, etc., to infer a set of complex mental states. *Theory of mind* refers to those social skills that allow humans to correctly attribute beliefs, goals, perceptions, feelings, and desires to the self and to others (Baron-Cohen, 1995; Leslie, 1994). Other sophisticated mechanisms such as *empathy* are used to understand the emotional and subjective states of others. These capabilities allow people to understand, explain, and predict the social behavior of others, and to respond appropriately.

To emulate human social perception, a robot must be able to identify who the person is (identification), what the person is doing (recognition), and how the person is doing it (emotive expression). Such information could be used by the robot to treat the person as an individual, to understand the person's surface behavior, and to potentially infer something about the person's internal states (e.g., the intent or the emotive state). Currently, there are vision-based systems capable of identifying faces, measuring head pose and gaze direction, recognizing gestures, and reading facial expressions. In the auditory domain, speech recognition and speaker identification are well-researched topics, and there is a growing interest in perceiving emotion in speech. New techniques and sensing technologies continue to be developed, becoming increasingly transparent to the user and perceiving a broader repertoire of human communication behavior. Not surprisingly, much of Kismet's perceptual system is specialized for perceiving and responding to people.

For robots to be human-aware, technologies for sensing and perceiving human behavior must be complemented with social cognition capabilities for understanding this behavior in social terms. As mentioned previously, humans employ theory-of-mind and empathy to infer and to reflect upon the intents, beliefs, desires, and feelings of others. In the field of narrative psychology, Bruner (1991) argues that stories are the most efficient and natural human way to communicate about personal and social matters. Schank & Abelson (1977) hypothesize that stories about one's own experiences and those of others (in addition to how these stories are constructed, interpreted, and interrelated) form the basic constituents of human memory, knowledge, social communication, self understanding, and the understanding of others. If robots shared comparable abilities with people to represent, infer, and reason about social behavior in familiar terms, then the communication and understanding of social behavior between humans and robots could be facilitated.

There are a variety of approaches to computationally understanding social behavior. Scassellati (2000a) takes a developmental psychology approach, combining two popular theories on the development of theory of mind in children (that of Baron-Cohen [1995] and Leslie [1994]), and implementing the synthesized model on a humanoid robot. In the tradition of AI reasoning systems, the BDI approach of Kinny et al. (1996) explicitly and symbolically models social expertise where agents attribute beliefs, desires, intents, abilities, and other mental states to others. In contrast, Schank & Abelson (1977) argue in favor of a story-based approach for representing and understanding social knowledge, communication, memory, and experience. Dautenhahn (1997) proposes a more embodied and interactive approach to understanding persons where storytelling (to tell autobiographic stories about oneself and to reconstruct biographic stories about others) is linked to the empathic, experiential way to relate other persons to oneself.

Being Understood

For a sociable robot to establish and maintain relationships with humans on an individual basis, the robot must understand people, and people should be able to intuitively understand the robot as they would others. It is also important for the robot to *understand its own self,* so that it can socially reason about itself in relation to others. Hence, in a similar spirit to the previous section, the same social skills and representations that might be used to understand others potentially also could be used by a robot understand its own internal states in social terms. This might correspond to possessing a theory-of-mind competence so that the robot can reflect upon its own intents, desires, beliefs, and emotions (Baron-Cohen, 1995). Such a capacity could be complemented by a story-based ability to construct, maintain, communicate about, and reflect upon itself and past experiences. As argued by Nelson (1993), autobiographical memory encodes a person's life history and plays an important role in defining the self.

Earlier, the importance of believability in robot design was discussed. Another important and related aspect is *readability*. Specifically, the robot's behavior and manner of expression (facial expressions, shifts of gaze and posture, gestures, actions, etc.) must be well matched to how the human observer intuitively interprets the robot's cues and movements to understand and predict its behavior (e.g., their theory-of-mind and empathy competencies). The human engaging the robot will tend to anthropomorphize it to make its behavior familiar and understandable. For this to be an effective strategy for inferring the robot's "mental states," the robot's outwardly observable behavior must serve as an accurate window to its underlying computational processes, and these in turn must be well matched to the person's social interpretations and expectations. If this match is close enough, the human can intuitively understand how to interact with the robot appropriately. Thus, readability supports the human's social abilities for understanding others. For this reason, Kismet has been designed to be a readable robot.

More demands are placed on the readability of robots as the social scenarios become more complex, unconstrained, and/or interactive. For instance, readability is reduced to believability in the case of passively viewed, non-interactive media such as classical animation. Here, observable behaviors and expressions must be familiar and understandable to a human observer, but there is no need for them to have any relation to the character's internal states. In this particular case, the behaviors are pre-scripted by animation artists, so there are no internal states that govern their behavior. In contrast, interactive digital pets (such as PF Magic's Petz or Bandai's Tamagotchi) present a more demanding scenario. People can interact with these digital pets within their virtual world via keyboard, mouse, buttons, etc. Although still quite limited, the behavior and expression of these digital pets is produced by a combination of pre-animated segments and internal states that determine which of these segments should be displayed. Generally speaking, the observed behavior is familiar and appealing to people if an intuitive relationship is maintained for how these states change with time, how the human can influence them, and how they are subsequently expressed through animation. If done well, people find these artifacts to be interesting and engaging and tend to form simple relationships with them.

Socially Situated Learning

For a robot, many social pressures demand that it continuously learn about itself, those it interacts with, and its environment. For instance, new experiences would continually shape the robot's personal history and influence its relationship with others. New skills and competencies could be acquired from others, either humans or other agents (robotic or otherwise). Hence, as with humans, robots must also be able to learn throughout their lifetime. Much of the inspiration behind Kismet's design comes from the socially situated learning and social development of human infants.

Many different learning strategies are observed in other social species, such as learning by imitation, goal emulation, mimicry, or observational conditioning (Galef, 1988). Some of these forms of social learning have been explored in robotic and software agents. For instance, learning by imitation or mimicry is a popular strategy being explored in humanoid robotics to transfer new skills to a robot through human demonstration (Schaal, 1997) or to acquire a simple proto-language (Billard & Dautenhahn, 2000). Others have explored social-learning scenarios where a robot learns about its environment by following around another robot (the model) that is already familiar with the environment. Billard and Dautenhahn (1998) show how robots can be used in this scenario to acquire a proto-language to describe significant terrain features.

In a more human-style manner, a robot could learn through tutelage from a human instructor. In general, it would be advantageous for a robot to learn from people in a manner that is natural for people to instruct. People use many different social cues and skills to help others learn. Ideally, a robot could leverage these same cues to foster its learning. In the next chapter, I explore in depth the question of learning from people as applied to humanoid robots.

1.4   Book Overview

This section offers a road map to the rest of the book, wherein I present the inspiration, the design issues, the framework, and the implementation of Kismet. In keeping with the infant-caregiver metaphor, Kismet's interaction with humans is dynamic, physical, expressive, and social. Much of this book is concerned with supplying the infrastructure to support socially situated learning between a robot infant and its human caregiver. Hence, I take care in each chapter to emphasize the constraints that interacting with a human imposes on the design of each system, and tie these issues back to supporting socially situated learning.

The chapters are written to be self-contained, each describing a different aspect of Kismet's design. It should be noted, however, that there is no central control. Instead, Kismet's coherent behavior and its personality emerge from all these systems acting in concert. The interaction between these systems is as important as the design of each individual system.

Evaluation studies with naive subjects are presented in many of the chapters to socially ground Kismet's behavior in interacting with people. Using the data from these studies, I evaluate the work with respect to the performance of the human-robot system as a whole.

• *Chapter 2* I motivate the realization of sociable robots and situate this work with Kismet with respect to other research efforts. I provide an in-depth discussion of socially situated learning for humanoid robots to motivate Kismet's design.

• *Chapter 3* I highlight some key insights from developmental psychology. These concepts have had a profound impact on the types of capabilities and interactions I have tried to achieve with Kismet.

• *Chapter 4* I present an overview of the key design issues for sociable robots, an overview of Kismet's system architecture, and a set of evaluation criteria.

• *Chapter 5* I describe the system hardware including the physical robot, its sensory configuration, and the computational platform. I also give an overview of Kismet's low-level visual and auditory perceptions. A detailed presentation of the visual and auditory systems follows in later chapters.

• *Chapter 6* I offer a detailed presentation of Kismet's visual attention system.

• *Chapter 7* I present an in-depth description of Kismet's ability to recognize affective intent from the human caregiver's voice.

• *Chapter 8* I give a detailed presentation of Kismet's motivation system, consisting of both homeostatic regulatory mechanisms as well as models of emotive responses. This system serves to motivate Kismet's behavior to maintain Kismet's internal state of "well-being."

• *Chapter 9* Kismet has several time-varying motivations and a broad repertoire of behavioral strategies to satiate them. This chapter presents Kismet's behavior system that arbitrates among these competing behaviors to establish the current goal of the robot. Given the goal of the robot, the motor systems are responsible for controlling Kismet's output modalities (body, face, and voice) to carry out the task. This chapter also presents an overview of

Kismet's diverse motor systems and the different levels of control that produce Kismet's observable behavior.

• *Chapter 10* I present an in-depth look at the motor system that controls Kismet's face. It must accommodate various functions such as emotive facial expression, communicative facial displays, and facial animation to accommodate speech.

• *Chapter 11* I describe Kismet's expressive vocalization system and lip synchronization abilities.

• *Chapter 12* I offer a multi-level view of Kismet's visual behavior, from low-level oculo-motor control to using gaze direction as a powerful social cue.

• *Chapter 13* I summarize our results, highlight key contributions, and present future work for Kismet. I then look beyond Kismet and offer a set of grand challenge problems for building sociable robots of the future.

1.5 Summary

In this chapter, I outlined the vision of sociable robots. I presented a number of well-known examples from science fiction that epitomize the vision of a sociable robot. I argued in favor of constructing such machines from the scientific pursuit of modeling and understanding social intelligence through the construction of a socially intelligent robot. From a practical perspective, socially intelligent technologies allow untrained human users to interact with robots in a way that is natural and intuitive. I offered a few applications (in the present, the near future, and the more distant future) that motivate the development of robots that can interact with people in a rich and enjoyable manner. A few key aspects of human social intelligence were characterized to derive a list of core ingredients for sociable robots. Finally, I offered Kismet as a detailed case study of a sociable robot for the remainder of the book. Kismet explores several (certainly not all) of the core ingredients, although many other researchers are exploring others.

2 Robot in Society: A Question of Interface

As robots take on an increasingly ubiquitous role in society, they must be easy for the average person to use and interact with. They must also appeal to different ages, genders, incomes, educations, and so forth. This raises the important question of how to properly interface untrained humans with these sophisticated technologies in a manner that is intuitive, efficient, and enjoyable to use. What might such an interface look like?

2.1 Lessons from Human Computer Interaction

In the field of human computer interaction (HCI), researchers are already examining how people interact with one form of interactive technology—computers. Recent research by Reeves and Nass (1996) has shown that humans (whether computer experts, lay-people, or computer critics) generally treat computers as they might treat other people. They treat computers with politeness usually reserved for humans. They are careful not to hurt the computer's "feelings" by criticizing it. They feel good if the computer compliments them. In team play, they are even are willing to side with a computer against another human if the human belongs to a different team. If asked before the respective experiment if they could imagine treating a computer like a person, they strongly deny it. Even after the experiment, they insist that they treated the computer as a machine. They do not realize that they treated it as a peer.

In these experiments, why do people unconsciously treat the computers in a social manner? To explain this behavior, Reeves and Nass appeal to evolution. Their main thesis is that the *human brain evolved in a world in which only humans exhibited rich social behaviors, and a world in which all perceived objects were real physical objects. Anything that seemed to be a real person or place was real* (Reeves & Nass, 1996, page 12). Evolution has hardwired the human brain with innate mechanisms that enable people to interact in a social manner with others that also behave socially. In short, we have evolved to be experts in social interaction. Although our brains have changed very little over thousands of years, we have to deal with modern technology. As a result, if a technology behaves in a socially competent manner, we evoke our evolved social machinery to interact with it. Reeves and Nass argue that it actually takes *more* effort for people to consciously inhibit their social machinery in order to *not* treat the machine in this way. From their numerous studies, they argue that a social interface may be a truly universal interface (Reeves & Nass, 1996).

From these findings, I take as a working assumption that technological attempts to foster human-technology relationships will be accepted by a majority of people *if* the technological gadget displays rich social behavior. Similarity of morphology and sensing modalities makes humanoid robots one form of technology particularly well-suited to this.

Sociable robots offer an intriguing alternative to the way humans interact with robots today. If the findings of Reeves and Nass hold true for humanoid robots, then those that participate in rich human-style social exchange with their users offer a number of advantages. First, people would find working with them more enjoyable and would thus feel more competent. Second, communicating with them would not require any additional training since humans are already experts in social interaction. Third, if the robot could engage in various forms of social learning (imitation, emulation, tutelage, etc.), it would be easier for the user to teach new tasks. Ideally, the user could teach the robot just as one would teach another person.

Hence, one important challenge is not only to build a robot that is an effective learner, but also to build a robot that can learn in a way that is natural and intuitive for people to teach. The human learning environment is a dramatically different learning environment from that of typical autonomous robots. It is an environment that affords a uniquely rich learning potential. Any robot that co-exists with people as part of their daily lives must be able to learn and adapt to new experiences using social interaction. As designers, we simply cannot predict all the possible scenarios that such a robot will encounter. Fortunately, there are many advantages social cues and skills could offer robots that learn from people (Breazeal & Scassellati, 2002).

I am particularly interested in the human form of socially situated learning. From Kismet's inception, the design has been driven by the desire to leverage from the social interactions that transpire between a robot infant and its human caregiver. Much of this book is concerned with supplying the infrastructure to support this style of learning and its many advantages. The learning itself, however, is the topic of future work.

2.2   Socially Situated Learning

Humans (and other animals) acquire new skills socially through direct tutelage, observational conditioning, goal emulation, imitation, and other methods (Galef, 1988; Hauser, 1996). These social learning skills provide a powerful mechanism for an observer (the learner) to acquire behaviors and knowledge from a skilled individual (the instructor). In particular, imitation is a significant social-learning mechanism that has received a great deal of interest from researchers in the fields of animal behavior and child development.

Similarly, social interaction can be a powerful way for transferring important skills, tasks, and information to a robot. A socially competent robot could take advantage of the same sorts of social learning and teaching scenarios that humans readily use. From an engineering perspective, a robot that could imitate the actions of a human would provide a simple and effective means for the human to specify a task and for the robot to acquire new skills without any additional programming. From a computer science perspective, imitation and other forms of social learning provide a means for biasing interaction and constraining the search space for learning. From a developmental psychology perspective, building systems that learn from humans allows us to investigate a minimal set of competencies necessary for social learning.

By positing the presence of a human that is motivated to help the robot learn the task at hand, a powerful set of constraints can be introduced to the learning problem. A good teacher is very perceptive to the limitations of the learner and scales the instruction accordingly. As the learner's performance improves, the instructor incrementally increases the complexity of the task. In this way, the learner is competent but slightly challenged—a condition amenable to successful learning. This type of learning environment captures key aspects of the learning environment of human infants, who constantly benefit from the help and encouragement of their caregivers. An analogous approach could facilitate a robot's ability to acquire more complex tasks in more complex environments. Keeping this goal in mind, outlined below are three key challenges of robot learning, and how social interaction can be used to address them in interesting ways (Breazeal & Scassellati, 2002).

Knowing What Matters

Faced with an incoming stream of sensory data, a robot (the learner) must figure out which of its myriad of perceptions are relevant to learning the task. As the perceptual abilities of a robot increase, the search space becomes enormous. If the robot could narrow in on those few relevant perceptions, the learning problem would become significantly more manageable.

Knowing what matters when learning a task is fundamentally a problem of determining saliency. Objects can gain saliency (that is, they become the target of attention) through a variety of means. At times, objects are salient because of their inherent properties; objects that move quickly, objects that have bright colors, and objects that are shaped like faces are all likely to attract attention. We call these properties *inherent* rather than *intrinsic* because they are perceptual properties, and thus are observer-dependent rather than a quality of an external object. Objects become salient through contextual effects. The current motivational state, emotional state, and knowledge of the learner can impact saliency. For example, when the learner is hungry, images of food will have higher saliency than otherwise. Objects can also become salient if they are the focus of the instructor's attention. For example, if the human is staring intently at a specific object, that object may become a salient part of the scene even if it is otherwise uninteresting. People naturally attend to the key aspects of a task while performing that task. By directing the robot's own attention to the object of the instructor's attention, the robot would automatically attend to the critical aspects of the task. Hence, a human instructor could indicate what features the robot should attend to as it learns how to perform the task. Also, in the case of social instruction, the robot's gaze direction could serve as an important feedback signal for the instructor.

Knowing What Action to Try

Once the robot has identified salient aspects of the scene, how does it determine what actions it should take? As robots become more complex, their repertoire of possible actions increases. This also contributes to a large search space. If the robot had a way of focusing on those potentially successful actions, the learning problem would be simplified.

In this case, a human instructor, sharing a similar morphology with the robot, could provide considerable assistance by demonstrating the appropriate actions to try. The body mapping problem is challenging, but could provide the robot with a good first attempt. The similarity in morphology between human and humanoid robot could also make it easier and more intuitive for the instructor to correct the robot's errors.

Instructional Feedback

Once a robot can observe an action and attempt to perform it, how can the robot determine whether or not it has been successful? Further, if the robot has been unsuccessful, how does it determine which parts of its performance were inadequate? The robot must be able to identify the desired outcome and to judge how its performance compares to that outcome. In many situations, this evaluation depends on understanding the goals and intentions of the instructor as well as the robot's own internal motivations. Additionally, the robot must be able to diagnose its errors in order to incrementally improve performance.

The human instructor, however, has a good understanding of the task and knows how to evaluate the robot's success and progress. If the instructor could communicate this information to the robot in a way that the robot could use, the robot could bootstrap from the instructor's evaluation in order to shape its behavior. One way a human instructor could facilitate the robot's evaluation process is by providing expressive feedback. The robot could use this feedback to recognize success and to correct failures. In the case of social instruction, the difficulty of obtaining success criteria can be simplified by exploiting the natural structure of social interactions. As the learner acts, the facial expressions (smiles or frowns), vocalizations, gestures (nodding or shaking of the head), and other actions of the instructor all provide feedback that allows the learner to determine whether it has achieved the goal.

In addition, as the instructor takes a turn, the instructor often looks to the learner's face to determine whether the learner appears confused or understands what is being demonstrated. The expressive displays of a robot could be used by the instructor to control the rate of information exchange—to either speed it up, to slow it down, or to elaborate as appropriate. If the learner appears confused, the instructor can slow down the training scenario until the learner is ready to proceed. Facial expressions could be an important cue for the instructor as well as the robot. By regulating the interaction, the instructor could establish an appropriate learning environment and provide better quality instruction.

Finally, the structure of instructional situations is iterative: the instructor demonstrates, the student performs, and then the instructor demonstrates again, often exaggerating or focusing on aspects of the task that were not performed successfully. The ability to take turns lends significant structure to the learning episode. The instructor continually modifies the way he/she performs the task, perhaps exaggerating those aspects that the student performed inadequately, in an effort to refine the student's subsequent performance. By repeatedly responding to the same social cues that initially allowed the learner to understand and identify the salient aspects of the scene, the learner can incrementally refine its approximation of the actions of the instructor.

For the reasons discussed above, many social-learning abilities have been implemented on Kismet. These include the ability to direct the robot's attention to establish shared reference, the ability for the robot to recognize expressive feedback such as praise and prohibition, the ability to give expressive feedback to the human, and the ability to take turns to structure the learning episodes. Chapter 3 illustrates strong parallels in how human caregivers assist their infant's learning through similar social interactions.

2.3   Embodied Systems That Interact with Humans

Before I launch into the presentation of my work with Kismet, I will summarize some related work. These diverse implementations overlap a variety of issues and challenges that my colleagues and I have had to overcome in building Kismet.

There are a number of systems from different fields of research that are designed to interact with people. Many of these systems target different application domains such as computer interfaces, Web agents, synthetic characters for entertainment, or robots for physical labor. In general, these systems can be either embodied (the human interacts with a robot or an animated avatar) or disembodied (the human interacts through speech or text entered at a keyboard). The embodied systems have the advantage of sending para-linguistic communication signals to a person, such as gesture, facial expression, intonation, gaze direction, or body posture. These embodied and expressive cues can be used to complement or enhance the agent's message. At times, para-linguistic cues carry the message on their own, such as emotive facial expressions or gestures. Cassell (1999b) presents a good overview of how embodiment can be used by avatars to enhance conversational discourse (there are, however, a number of systems that interact with people without using natural language). Further, these embodied systems must also address the issue of sensing the human, often focusing on perceiving the human's embodied social cues. Hence, the perceptual problem for these systems is more challenging than that of disembodied systems. In this section I summarize a few of the embodied efforts, as they are the most closely related to Kismet.

Embodied Conversation Agents

There are a number of graphics-based systems that combine natural language with an embodied avatar (see figure 2.1 for a couple of examples). The focus is on natural, conversational discourse accompanied by gesture, facial expression, and so forth. The human uses these systems to perform a task, or even to learn how to perform a task. Sometimes, the task could simply be to communicate with others in a virtual space, a sort of animated "chatroom" with embodied avatars (Vilhjalmsson & Cassell, 1998).

There are several fully embodied conversation agents under development at various institutions. One of the most advanced systems is Rea from the Media Lab at MIT (Cassell et al., 2000). Rea is a synthetic real-estate agent, situated in a virtual world, that people can query about buying property. The system communicates through speech, intonation, gaze direction, gesture, and facial expression. It senses the location of people in the room and recognizes a few simple gestures. Another advanced system is called Steve, under development at USC (Rickel & Johnson, 2000). Steve is a tutoring system, where the human is immersed in virtual reality to interact with the avatar. It supports domain-independent capabilities to support task-oriented dialogs in 3D virtual worlds. For instance, Steve trains people how to operate a variety of equipment on a virtual ship and guides them through the ship to show them where the equipment is located. Cosmo, under development at North Carolina State University, is an animated Web-based pedagogical agent for children (Lester et al., 2000). The character inhabits the Internet Advisor, a learning environment for the domain of Internet packet routing. Because the character interacts with children, particular

Figure 2.1
Some examples of embodied conversation agents. To the left is Rea, a synthetic real estate agent. To the right is BodyChat, a system where online users interact via embodied animated avatars. Images courtesy of Justine Cassell and Hannes Vilhjálmsson of the Gesture and Narrative Language Research Group. Images © MIT Media Lab.

attention is paid to the issues of life-like behavior and engaging the students at an affective level.

There are a number of graphical systems where the avatar predominantly consists of a face with minimal to no body. A good example is Gandalf, a precursor system of Rea. The graphical component of the agent consisted of a face and a hand. It could answer a variety of questions about the solar system but required the user to wear a substantial amount of equipment in order to sense the user's gestures and head orientation (Thorisson, 1998). In Takeuchi and Nagao (1993), the use of an expressive graphical face to accompany dialogue is explored. They found that the facial component was good for initiating new users to the system, but its benefit was not as pronounced over time.

Interactive Characters

There are a variety of interactive characters under development for the entertainment domain. The emphasis for each system is compelling, life-like behavior and characters with personality. Expressive, readable behavior is of extreme importance for the human to understand the interactive story line. Instead of passively viewing a scripted story, the user creates the story interactively with the characters.

A number of systems have been developed by at the MIT Media Lab (see figure 2.2). One of the earliest systems was the ALIVE project (Maes et al., 1996). The best-known character of this project is Silas, an animated dog that the user could interact with using gesture within a virtual space (Blumberg, 1996). Several other systems have since been

Figure 2.2
Some examples of life-like characters. To the left are the animated characters of *Swamped!*. The racoon is completely autonomous, whereas the human controls the animated chicken through a plush toy interface. To the right is a human interacting with *Silas* from the ALIVE project. Images courtesy of Bruce Blumberg from the Synthetic Characters Group. Images © MIT Media Lab.

developed at the Media Lab by the Synthetic Characters Group. For instance, in *Swamped!* the human interacts with the characters using a sensor-laden plush chicken (Johnson et al., 1999). By interacting with the plush toy, the user could control the behavior of an animated chicken in the virtual world, which would then interact with other characters.

There are several synthetic character systems that support the use of natural language. The Oz project at CMU is a good example (Bates, 1994). The system stressed "broad and shallow" architectures, biasing the preference for characters with a broad repertoire of behaviors over those that are narrow experts. Some of the characters were graphics-oriented (such as woggles), whereas others were text-based (such as Leotard the cat). Using a text-based interface, Bates et al. (1992) explored the development of social and emotional agents. At Microsoft Research Labs, Peedy was an animated parrot that users could interact with in the domain of music (Ball et al., 1997). In later work at Microsoft Research, Ball and Breese (2000) explored incorporating emotion and personality into conversation agents using a Baysian network technique.

Human-Friendly Humanoids

In the robotics community, there is a growing interest in building personal robots, or in building robots that share the same workspace with humans. Some projects focus on more advanced forms of tele-operation. Since my emphasis is on autonomous robots, I will not dwell on these systems. Instead, I concentrate on those efforts in building robots that interact with people.

There are several projects that focus on the development of robot faces (a few examples are shown in figure 2.3). For instance, researchers at the Science University of Tokyo have developed human-like robotic faces (typically resembling a Japanese woman) that

Figure 2.3
Some examples of faces for humanoid robots. To the left is a very human-like robot developed at the Science University of Tokyo. A robot more in the spirit of a mechanical cartoon (developed at Waseda University) is shown in the middle picture. To the right is a stylized but featureless face typical of many humanoid robots (developed by the Kitano Symbiotic Systems Project).

incorporate hair, teeth, silicone skin, and a large number of control points (Hara, 1998). Each control point maps to a facial action unit of a human face. The facial action units characterize how each facial muscle (or combination of facial muscles) adjust the skin and facial features to produce human expressions and facial movements (Ekman & Friesen, 1982). Using a camera mounted in the left eyeball, the robot can recognize and produce a predefined set of emotive facial expressions (corresponding to anger, fear, disgust, happiness, sorrow, and surprise). A number of simpler expressive faces have been developed at Waseda University, one of which can adjust its amount of eye-opening and neck posture in response to light intensity (Takanobu et al., 1999).

The number of humanoid robotic projects under way is growing, with a particularly strong program in Japan (see figure 2.4). Some humanoid efforts focus on more traditional challenges of robot control. Honda's P3 is a bipedal walker with an impressive human-like gait (Hirai, 1998). Another full-bodied (but non-locomotory) humanoid is at ATR (Schaal, 1999). Here, the focus has been on arm control and in integrating arm control with vision to mimic the gestures and tasks demonstrated by a human. There are several upper-torso humanoid robots. NASA is developing a humanoid robot called Robonaut that works with astronauts to perform a variety of tasks while in orbit, such as carrying out repairs on the external surface of the space shuttle (Ambrose et al., 1999). One of the most well-known humanoid robots is Cog, under development at the MIT Artificial Intelligence Lab (Brooks et al., 1999). Cog is a general-purpose humanoid platform used to explore theories and models of intelligent behavior and learning, both physical and social.

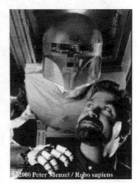

Figure 2.4
Some examples of humanoid robots. To the left is Cog, developed at the MIT AI Lab. The center picture shows Honda's bipedal walking robot, P3. The right picture shows NASA's Robonaut.

Personal Robots

There are a number of robotic projects that focus on operating within human environments. Typically these robots are not humanoid in form, but are designed to support natural communication channels such as gesture or speech.

There are a few robots that are being designed for domestic use. For systems such as these, safety and minimizing impact on human living spaces are important issues as well as performance and ease of use. Many applications of this kind focus on providing assistance to the elderly or to the disabled. The MOVAID system (Dario & Susani, 1996) and a similar project at Vanderbilt University (Kawamura et al., 1996). In a somewhat related effort, Dautenhahn (1999) has employed autonomous robots to assist in social therapy of fairly high-functioning autistic children.

In the entertainment market, there are a growing number of synthetic pets (both robotic and digital). Sony's robot dog Aibo (shown in figure 2.5) can perceive a few simple visual and auditory features that allow it to interact with a pink ball and objects that appear skin-toned. It is mechanically quite sophisticated, able to locomote, to get up if it falls down, and to perform an assortment of tricks. There are simpler, less expensive robotic dogs such as Tiger Electronic's iCybie. One of the first digital pets include Tamagotchis which the child could carry with him/her on a keychain and care for (or the toy would get "sick" and eventually "die"). There are also animated pets that live on the computer screen such

©2000 Peter Menzel/Robo sapiens

Figure 2.5
Sony's Aibo is a sophisticated robot dog.

as PF Magic's Petz. Their design intentionally encourages people to establish a long-term relationship with them.

2.4 Summary

In this chapter, I have motivated the construction of sociable robots from the viewpoint of building robots that are natural and intuitive to communicate with and to teach. I summarized a variety of related efforts in building embodied tchnologies that interact with people. My work with Kismet is concerned both with supporting human-style communication as well as with providing the infrastructure to support socially situated learning. I discussed how social interaction and social cues can address some of the key challenges in robot learning in new and interesting ways. These are the capabilities I have taken particular interest in building into Kismet.

3 Insights from Developmental Psychology

Human babies become human beings because they are treated as if they already were human beings.
—J. Newson (1979, p. 208)

In this chapter, I discuss the role social interaction plays in learning during infant-caregiver exchanges. First, I illustrate how the human newborn is primed for social interaction immediately after birth. This fact alone suggests how critically important it is for the infant to establish a social bond with his caregiver, both for survival purposes as well as to ensure normal cognitive and social development. Next, I focus on the caregiver and discuss how she employs various social acts to foster her infant's development. I discuss how infants acquire meaningful communication acts through ongoing interaction with adults. I conclude this chapter by relating these lessons to Kismet's design.

The design of Kismet's synthetic nervous system is heavily inspired by the social development of human infants. This chapter illustrates strong parallels to the previous chapter in how social interaction with a benevolent caregiver can foster robot learning. By implementing similar capabilities as the initial perceptual and behavioral repertoire of human infants, I hope to prime Kismet for natural social exchanges with humans and for socially situated learning.

3.1 Early Infant-Caregiver Interactions

Immediately after birth, human infants are immersed in a dynamic and social world. A powerful bond is quickly formed between an infant and the caregiver who plays with him and nurtures him. Much of what the infant learns is acquired through this social scenario, in which the caregiver is highly socially sophisticated and culturally competent, whereas the infant is naive.

From birth, infants demonstrate a preference for humans over other forms of stimuli (Trevarthen, 1979). Certain types of spontaneous events can momentarily dominate the infant's attention (such as primary colors, movement, and sounds), but human-mediated events are particularly good at sustaining it. Humans certainly encompass a myriad of attention-getting cues that infants are biologically tuned to react to (coordinated movement, color, and so forth). However, infants demonstrate significant attention to a variety of human-specific stimuli. For instance, even neonates exhibit a preference for looking at simple face-like patters (Fantz, 1963). When looking at a face, infants seem particularly drawn to gazing at the eyes and mouth (Caron et al., 1973). Human speech is also particularly attractive, and infants show particular preference to the voices of their caregivers (Mills & Melhuish, 1974; Hauser, 1996). Brazelton (1979) discusses how infants are particularly attentive to human faces and softly spoken voices. They communicate this preference through attentive regard, a "softening" of their face and eyes, and a prolonged suppression of body movement. More significantly, however, humans respond contingently to an infant's own actions. Caregivers, in particular, frequently respond to an infant's immediately preceding actions. As a result, the infant is particularly responsive to his caregiver, and the caregiver is particularly good at acquiring and sustaining her infant's attention. According to Newson, "this simple contingent reactivity makes her an object of absolute, compelling interest to her baby" (Newson, 1979, p. 208).

Not only are infants born with a predisposition to respond to human social stimuli, they also seem biologically primed to respond in a recognizable social manner (Trevarthen, 1979). Namely, infants are born with a set of well-coordinated *proto-social responses* which allow them to attract and engage adults in rich social exchanges. For instance, Johnson (1993) argues that the combination of having a limited depth of field[1] with early fixation patterns forces the infant to look predominantly at his caregiver's face. This brings the infant into face-to-face contact with his caregiver, which encourages her to try to engage him socially. Trevarthen (1979) discusses how infants make *prespeech* movements with their lips and tongue, gives them the appearance of trying to respond with speech-like sounds. Kaye (1979) discusses a scenario where the burst-pause-burst pattern in suckling behavior, coupled with the caregiver's tendency to jiggle the infant during the pauses, lays the foundation of the earliest forms of turn-taking that becomes more flexible and regular over time. This leads to more fluid exchanges with the caregiver while also allowing her to add structure to her teaching scenarios with him. It is posited that infants engage their caregivers in imitative exchanges, such as mirroring facial expressions (Meltzoff & Moore, 1977) or the pitch and duration of sounds (Maratos, 1973). Trevarthen (1979) discusses how the wide variety of facial expressions displayed by infants are interpreted by the caregiver as indications of the infant's motivational state. They serve as his responses to her efforts to engage him, and she uses them as feedback to carry the "dialogue" along.

Together, the infant's biological attraction to human-mediated events in conjunction with his proto-social responses serve to launch him into social interactions with his caregiver. There is an imbalance, however, in the social and cultural sophistication of the two partners. Fortunately, there are a number of ways in which an infant limits the complexity of his interactions with the world. This is a critical skill for social learning because it allows the infant to keep himself from being overwhelmed or under-stimulated for prolonged periods of time. Tronick et al. (1979) note that this mismatch is critical for the infant's development because it provides more and more complicated events to learn about. Generally speaking,

---

1. A newborn's resolution is restricted to objects approximately 20 cm away, about the distance to his caregiver's face when she holds him.

as the infant's capabilities improve and become more diverse, there is still an environment of sufficient complexity for him to develop into.

For instance, the infant's own physically immature state serves to limit his perceptual and motor abilities, which simplifies his interaction with the world. According to Tronick et al. (1979), infants perceive events within a narrower peripheral field and a shorter straight-ahead space than adults and older children. Further, the infant's inability to distinguish separate words in his caregiver's vocalizations may allow him to treat her complex articulated phrases as being similar to his own simpler sounds (Bateson, 1979; Trehub & Trainor, 1990). This allows the infant to participate in *proto-dialogues* with her, from which he can begin to learn the tempo, intonation, and emotional content of language long before speaking and understanding his first words (Fernald, 1984). In addition, the infant is born with a number of innate behavioral responses that constrain the sorts of stimulation that can impinge upon him. Various reflexes (such as quickly withdrawing his hand from a painful stimulus, evoking the looming reflex in response to a quickly approaching object, and closing his eyelids in response to a bright light) serve to protect the infant from stimuli that are potentially dangerous or too intense. According to Brazelton (1979), when the infant is in a situation where his environment contains too much commotion and confusing stimuli, he either cries or tightly shuts his eyes. By doing so, he shuts out the disturbing stimulation.

To assist the caregiver in regulating the intensity of interaction, the infant provides her with cues as to whether he is being under-stimulated or overwhelmed. When the infant feels comfortable in his surroundings, he generally appears content and alert. Too much commotion results in an appearance of anxiety, or crying, if the caregiver does not act to correct the environment. In contrast, too much repetition causes habituation or boredom (often signaled by the infant looking away from the stimulus). For the caregiver, the ability to present an appropriately complex view of the world to her infant strongly depends on how good she is at reading her infant's expressive and behavioral cues.

Adults naturally engage infants in appropriate interactions without realizing it, and care-givers seem to be instinctually biased to do so, varying the rate, intensity, and quality of their activities from that of adult-to-adult exchanges. Tronick et al. (1979) state that just about everything the caregiver does is exaggerated and slowed down. *Parentese* (or *motherese*) is a well-known example of how adults simplify and exaggerate important aspects of language such as pitch, syntax, and pronunciation (Bateson, 1979; Hirsh-Pasek et al., 1987). By doing so, adults may draw the infant's attention to salient features of the adult's vocalizations and hold the infant's attention (Fernald, 1984). During playful exchanges, caregivers are quite good at bringing their face sufficiently close to their infant, orienting straight ahead, being careful to move either parallel or perpendicular to the infant, and using exaggerated facial expressions to make the face more readable for the infant's developing visual system.

3.2   Development of Communication and Meaning

It is essential for the infant's psychological development that her caregiver treat her as an intentional being. Both the infant's responses and her parent's own caregiving responses have been selected for because they foster this kind of interaction. This, in turn, serves to bootstrap the infant into a cultural world. Trevarthen (1979) argues that infants must exhibit *subjectivity* (i.e., the ability to clearly demonstrate to others by means of coordinated actions at least the rudiments of intentional behavior) to be able to engage in interpersonal communication. According to Newson (1979), the early proto-social responses exhibited by infants are a close enough approximation to the adult forms that the caregiver interprets his infant's reactions by a process of *adultomorphism*. Simply stated, he treats his infant as if she is already fully socially aware and responsive—with thoughts, wishes, intents, desires, and feelings that she is trying to communicate to him as any other person would. He credits his infant's actions (which may be spontaneous, reflexive, or accidental) with social significance and treats them as her attempt to carry out a meaningful dialogue with him. This allows him to impute meaning to the exchange in a consistent and reliable manner and to establish a dialogue with her. It is from these exchanges that the communication of shared meanings gradually begins to take form.

By six weeks, human infants and their caregivers are communicating extensively face-to-face. During nurturing or playful exchanges, the baby's actions include vocalizing, crying, displaying facial expressions, waving, kicking, satisfied sucking or snuggling, and so on, which the caregiver interprets as her attempts to communicate her thoughts, feelings, and intentions to him. At an infant's early age, Kaye (1979) and Newson (1979) point out that it is the caregiver who supplies the meaning to the exchange, and it is the proto-social skill of early turn-taking that allows him to maintain the illusion that a meaningful conversation is taking place. When his infant does something that can be interpreted as a turn in the proto-dialogue, he treats it as such. He fills the gaps with her responses and pauses to allow her to respond, while allowing himself to be paced by her but also gently encouraging her.

The pragmatics of conversation are established during these proto-dialogues which in turn plays an important role in how meaning emerges for the infant. Schaffer (1977) writes that turn-taking of the "non-specific, flexible, human variety" prepares the infant for several important social developments. First, it allows the infant to discover what sorts of activity on her part will get responses from her caregiver. Second, it allows routine, predictable sequences to be established that provide a context of mutual expectations. This is possible due to the caregiver's consistent and predictable manner of responding to his infant *because* he assumes that she is fully socially responsive and shares the same meanings that he applies to the interaction. Eventually, the infant exploits these consistencies to learn the significance her actions and expressions have for other people—to the point where she *does* share the same meanings.

Halliday (1975) explores the acquisition of meaningful communication acts from the viewpoint of how children *use* language to serve themselves in the course of daily life. He refers to the child's first language (appearing around six months of age) as a *proto-language,* which consists of the set of acquired meanings shared by infant and adult. During this phase, the infant is able to use her voice to influence the behavior of others (although in a manner that bears little resemblance to the adult language). Furthermore, she soon learns how to apply these meaningful vocal acts in appropriate and significant contexts. To paraphrase Halliday (1975, p. 11), the infant uses her voice to order people about, to get them to do things for her; she uses it to demand certain objects or services; she uses it to make contact with people, to feel close to them; and so on. All these things are meaningful actions. Hence, the baby's vocalizations hold meaning to both baby and adult long before she ever utters her first words (typically about a year later). All the while, caregivers participate in the development of the infant's proto-language by talking to the infant in a manner that she can interpret within her limitations, and at the same time gently pushing her understanding without going too far.

Siegel (1999) argues that, in a similar way, caregivers bootstrap their infant into performing intentional acts (i.e., acts *about* something) significantly before the infant is capable of true intentional thought. Around the age of four months, the infant is finally able to break her caregiver's gaze to look at other things in the world. The caregiver interprets this break of gaze as an intentional act where the infant is now attending to some other object. In fact, Collis (1979) points out that the infant's gaze does not seem to be directed at anything in particular, nor does she seem to be trying to tell her caregiver that she is interested in some object. Instead, it is the caregiver who then turns a particular object into the object of attention. For instance, if an infant makes a reach and grasping motion in the direction of a given object, he will assume that the infant is interested in that object and is trying to hold it. In response, he intervenes by giving the object to the infant, thereby "completing" the infant's action. By providing this supporting action, he has converted an arbitrary act on the part of the infant into an action *about* something, thereby giving the infant's action intentional significance. In time, the infant begins to learn the consequences of her actions, and she begins to perform them with intent. Before this, however, the caregiver provides her with valuable experience by assisting her in behaving in an intentional manner.

3.3   Scaffolding for Social Learning

It is commonplace to say that caregiver-infant interaction is mutually engaging, where each partner adapts to the other over time. However, each has a distinctive role in the dyad—they are not equal partners. Tronick et al. (1979) liken the interaction between caregiver and infant to a duet played by a maestro and inept pupil (where the pupil is only seemingly dominant). The maestro continually makes adjustments to add variety and richness to the interplay, while allowing the pupil to participate in, experience, and learn from a higher level of performance than the pupil could accomplish on his own. Indeed, the caregiver's role is targeted toward developing the social sophistication of her infant to approach her own.

As traditionally viewed by the field of developmental psychology, *scaffolding* is conceptualized as a supportive structure provided by an adult whereby the adult manipulates the infant's interactions with the environment to foster novel abilities (Wood et al., 1976). Commonly viewed in social terms, it involves reducing distractions, marking the task's critical attributes, giving the infant affective forms of feedback, reducing the number of degrees of freedom in the target task, enabling the infant to experience the desired outcome before he is cognitively or physically able of seeking and attaining it for himself, and so forth. For instance, by exploiting the infant's instinct to perform a walking motion when supported upright, parents encourage their infant to learn how to walk before he is physically able. In this view, scaffolding is used as a pedagogical device where the adult provides deliberate support and guidance to push the infant a little beyond his current abilities to enable him to learn new skills.

Another notion of scaffolding stresses the importance of proto-social responses and their ability to bootstrap infants into social interactions with their caregivers. This form of scaffolding is referred to as *emergent scaffolding* by Hendriks-Jansen (1996). Here the caregiver-infant dyad is seen as two tightly coupled dynamic systems. In contrast to the previous case where the adult deliberately guides the infant's behavior to a desired outcome, instead the interaction is more free-form and arises from the continuous mutual adjustments between the two participants. For instance, the interaction between a suckling infant and the caregiver who jiggles him whenever he pauses in feeding creates a recognizable pattern of interaction. This interaction pattern encourages the habit of turn-taking, the importance of which was discussed earlier. Many of these early action patterns that newborns exhibit (such as this burst-pause-burst suckling pattern) have no place in adult behavior. They simply serve a bootstrapping role to launch the infant into the socio-cultural environment of adults, where important skills can then be transferred from adult to child.

Looking within the infant, there is a third form of scaffolding. For the purposes here, I call it *internal scaffolding*. This internal aspect refers to the incremental construction of the cognitive structures themselves that underlie observable behavior. Here, the form of the more mature cognitive structures are bootstrapped from earlier forms. Because these earlier forms provide the infant with some level of competence in the world, they are a good starting point for the later competencies to improve upon. In this way, the earlier structures foster and facilitate the learning of more sophisticated capabilities.

Hence, the infant is socially and culturally naive as compared to his caregiver. However, he is born with a rich set of well-coordinated proto-social responses that elicit nurturing, playful, and instructive behaviors from his caregiver. Furthermore, they encourage the caregiver to treat him as being fully socially responsive, sharing the same interpretation of the events that transpire during the interaction as she does. This imposes consistency on her responses to him, which is critical for learning. She plays the maestro in the caregiver-infant duet, providing various forms of scaffolding, in order to enhance and complement her infant's responses and to prolong the "performance" as long as possible. As she tries to win her infant's attention and sustain his interest, she takes into account her infant's current level of psychological and physiological abilities, his level of arousal, and his attention span. Based on these considerations, she adjusts the timing of her responses, introduces variations on a common theme to the interaction, and tries to balance the infant's agenda with her own agenda for him (Kaye, 1979).

The way the caregiver provides this scaffolding reflects her superior level of sophistication over her infant, and she uses this expertise to coax and guide her infant down a viable developmental path. For the remainder of this section, I discuss the various forms that scaffolding can take during social exchange, and how these forms foster the infant's development.

Directing attention   Bateson (1979) argues that the learning rate of infants is accelerated during social exchanges because caregivers focus their infants' attention on what is important. As discussed earlier, infants are able to direct attention to salient stimuli (especially toward social stimuli) at a very early stage of development. The caregiver leverages her infant's innate perceptual predispositions to first initiate an exchange by getting his attention and then artfully directs his attention during the exchange to other objects and events (such as directing the interaction to be about a particular toy). If his attention wanes, she will try to re-engage him by making either herself or the toy more salient through introducing motion, moving closer toward him, assuming a staccato manner of speech, and so forth. This helps to sustain his attention and interest on the most salient aspects of the interaction that she would like him to learn from. Furthermore, by directing the infant's attention to a desired stimulus, the caregiver can establish *shared reference*. This is a key component of social modeling theory and generally facilitates the learning problem presented to the learner as argued by Pepperberg (1988).

Affective feedback   Caregivers provide expressive feedback to their infants in response to the situations and events that their infants encounter. These affective responses can serve as socially communicated reinforcers for the infant. They can also serve as an affective assessment of a novel situation that the infant uses to organize his own behavior. In *social referencing,* this assessment can occur via visual channels whereby the infant looks to the caregiver's face to see her own affective reaction to an unfamiliar situation (Siegel, 1999). The assessment can also be communicated via auditory channels whereby the prosodic exaggerations typical of infant-directed speech (especially when communicating praise, prohibition, soothing, or attentional bids) are particularly well-matched to the innate affective responses of human infants (Fernald, 1989). This allows a caregiver to readily use either his voice or face to cause the infant to either relax or become more vigilant in certain situations, and to either avoid or approach objects that may be unfamiliar (Fernald, 1993). Given the number of important and novel situations that human infants encounter (which do not result in immediate pain or act as some other innate reinforcer, such as food), expressive feedback plays an important role in their social and behavioral development.

Regulating arousal   In addition to influencing the infant's attention and affective state, a caregiver is also careful to regulate the infant's arousal level. She may adopt a staccato manner of speech or use larger, faster movements to arouse him. Conversely, she uses soothing vocalizations and slower, smoother movements to relax him. Maintaining an optimal level of arousal is important, since performance and learning depend upon the infant being suitably alert, attentive, and interested in the situation at hand. Indeed, a caregiver expends significant effort in keeping her infant at a moderate level of arousal, where he is neither under-stimulated nor overwhelmed by the events facing him (Kaye, 1979).

Balancing agendas   During instructional interaction, the caregiver allows her infant to take the lead but shapes his agenda to meet her own. To accomplish this, the caregiver often flashes to where her infant is, and then attempts to pull his behavior in the direction she wants him to go. This agenda-shaping process can be seen when a caregiver imitates her infant. This is not simply a matter of mimicry. Instead, the caregiver employs a number of imitative strategies to shape and direct her infant's behavior with respect to her own. Kaye (1979) identifies three distinct strategies. First, *maximizing imitation* further exaggerates the infant's behavior. For instance, if the baby opens his mouth, she will open her mouth in an exaggerated manner to encourage him to open his wider. Alternatively, she may employ *minimizing imitation* to lessen the infant's behavior. For example, if baby begins to make a cry face, she responds with a quick cry face that immediately changes to a happy expression. She may also employ *modulating imitation* to shape his behavior. For instance, when a baby whines "waaah," the caregiver responds with the same whine but then softens to a soothing "awwww." Hence, it is often the case that the caregiver's imitation of her infant is motivated by her agenda for him.

Introducing repetition and variation   The caregiver frequently repeats movements and vocalizations as she engages her infant, but she is also very creative in introducing variations about a theme. According to Stern (1975), repetitive presentations of this nature are optimal for holding the infant's attention and establishes a good learning environment for him. Sometimes she presents several nearly identical acts or vocalizations in a row, separated by short pauses of varying duration. At other times she presents a series of markedly different acts or vocalizations that occupy nearly identical time slots. This simplifies the complexity of the stimulus the infant encounters by holding many of the features fairly constant while only varying a small number. This also helps to make the caregiver's behavior more predictable for the infant.

Timing and contingency   During social interactions, the caregiver adjusts the timing of her responses to make her responses contingent upon those of her infant, and to make his responses seemingly contingent upon hers. To accomplish this, she is very aware of her infant's physiological and psychological limitations and carefully observes him to make adjustments in her behavior. For instance, when talking with her infant she fills his pauses with her own utterances or gestures, and purposely leaves spaces between her own repetitive utterances and gestures for him to fill (Newson, 1979). She intently watches and listens for new initiatives from him, and immediately pauses when she thinks that he is about to respond. By doing so, she tries to establish or prolong a run of alternations between herself and her infant, sustaining his interest, and trying to get him to respond contingently to her (Kaye, 1979). During the interchange, each partner's movements and vocalizations demonstrate strong synchronization both within their turn and even across turns (Collis, 1979). Namely, the infant *entrains* to the caregiver's speech and gestures, and vice versa. This helps to establish an overall rhythm to the interplay, making it smoother and more synchronized over time.

Establishing games   It is important that each caregiver and infant pair develop its own set of conventional games. To paraphrase Kaye (1979), these games serve as the foundation of future communication and language-learning skills. They establish the process of defining conventions and roles, set up a mutual topic-comment format, and impose consistency and predictability on dyadic routines. These ritualized structures assist the infant in learning how to anticipate when and how a partner's behavior will change. Much of the social experience the infant is exposed to comes in the form of games. In general, games serve as an important form of scaffolding for infants.

From these scaffolded interactions, the infant very quickly learns how to socially manipulate people who care about him and for him. For instance, he learns how to get their attention, to playfully engage them, and to elicit nurturing responses from them. This is possible because his caregiver's scaffolding acts continually allow him to experience a higher level of functioning than he could achieve on his own. As he learns the significance his actions have for others, these initiatives become more deliberate and intentional. He also gradually begins to take on a more equal role in the interaction. For instance, he begins to adjust his timing, imitate his caregiver, and so forth (Tronick et al., 1979). As noted by Kaye (1979, p. 204), "This in turn gives him even finer control over the adult's behavior, so that he gains further information and more and more models of motor skills, of communication, and eventually of language. By the time his representational and phonemic systems are ready to begin learning language, he is already able to make his intentions understood most of the time, to orient himself in order to read and interpret other's responses, to elicit repetitions and variations."

3.4   Proto-Social Responses for Kismet

Our goal is for people to interact, play, and teach Kismet as naturally as they would an infant or very young child. These interactions provide many different kinds of scaffolding that Kismet could potentially use to foster its own learning. As a prerequisite for these interactions, people need to ascribe precocious social intelligence to Kismet, much as caregivers do for their infants. In doing so, people will treat Kismet as a socially aware creature and provide those interactions that Kismet will need to learn to become socially sophisticated.

For people to treat Kismet as a socially aware being, it needs to convey subjective internal states: intents, beliefs, desires, and feelings. The robot can be designed to exploit our natural human tendencies to respond socially to certain behaviors. To accomplish this, my colleagues and I have implemented several infant-like social cues and responses that human infants exhibit.

Acts that make subjective processes overt include focusing attention on objects, orienting to external events, handling or exploring objects with interest, and so forth. Summarizing the discussions of this chapter, I divide these responses into four categories. These are listed below. By implementing these four classes of responses (*affective, exploratory, protective,* and *regulatory*), I aim to encourage a person to treat Kismet as a social creature and to establish meaningful communication with it.

• *Affective responses* allow the human to attribute feelings to the robot.

• *Exploratory responses* allow the human to attribute curiosity, interest, and desires to the robot, and can be used to direct the interaction toward objects and events in the world.

• *Protective responses* keep the robot away from damaging stimuli and elicit concerned and caring responses from the human.

• *Regulatory responses* maintain a suitable environment that is neither too overwhelming nor under-stimulating, and tunes the human's behavior in a natural and intuitive way to the competency of the robot.

Of course, once Kismet can partake in social interactions with people, it is also important that the *dynamics* of the interaction be natural and intuitive. For this, I take the work of

Tronick et al. (1979) as a guide. They identify five phases that characterize social exchanges between three-month-old infants and their caregivers: *initiation, mutual-orientation, greeting, play-dialogue,* and *disengagement.* Each phase represents a collection of behaviors that mark the state of the communication. Not every phase is present in every interaction. For example, a greeting does not ensue if mutual orientation is not established. Furthermore, a sequence of phases may appear multiple times within a given exchange, such as repeated greetings before the play-dialogue phase begins. This is discussed in depth in chapter 9.

Acquiring a genuine proto-language is beyond the scope of this book, but learning how to mean and how to communicate those meanings to another (through voice, face, body, etc.) is a fundamental capacity of a socially intelligent being. These capacities have profoundly motivated the creation of Kismet. Hence, what is conceptualized and implemented in this work is heavily inspired and motivated by the processes highlighted in this chapter. I have endeavored to develop a framework that could ultimately be extended to support the acquisition of a proto-language and this characteristically human social learning process.

3.5   Summary

There are several key insights to be gleaned from the discussion in this chapter. The first is that human infants are born ready for social interaction with their caregivers. The initial perceptual and behavioral responses bias an infant to interact with adults and encourage a caregiver to interact with and care for him. Specifically, many of these responses enable the caregiver to carry on a "dialogue" with him. Second, the caregiver uses scaffolding to establish a consistent and appropriately complicated social environment for the infant that he can predict, steer, and learn from. She allows him to act as if he is in charge of leading the dialogue, but she is actually the one in charge. By doing so, she allows the infant to experiment and learn how his responses influence her. Third, the development of the infant's acts of meaning is inherently a social process, and it is grounded in having the infant learn how he can use his voice to serve himself. It is important to consider the infant's motivations—why he is motivated to use language and for what reasons. These motivations drive what he learns and why. These insights have inspired the design of Kismet's synthetic nervous system—from the design of each system to the proto-social skills and abilities they implement. My goal is for people to play with Kismet as they would an infant, thereby providing those critical interactions that are needed to develop social intelligence and to become a social actor in the human world.

4 Designing Sociable Robots

The challenge of building Kismet lies in building a robot that is capable of engaging humans in natural social exchanges that adhere to the infant-caregiver metaphor. The motivation for this kind of interaction highlights my interest in social development and in socially situated learning for humanoid robots. Consequently, this work focuses on the problem of building the physical and computational infrastructure needed to support these sorts of interactions and learning scenarios. The social learning, however, is beyond the scope of this book.

Inspired by infant social development, psychology, ethology, and evolutionary perspectives, this work integrates theories and concepts from these diverse viewpoints to enable Kismet to enter into natural and intuitive social interaction with a human caregiver. For lack of a better metaphor, I refer to this infrastructure as the robot's *synthetic nervous system* (SNS).

4.1   Design Issues for Sociable Robots

Kismet is designed to perceive a variety of natural social cues from visual and auditory channels, and to deliver social signals to the human caregiver through gaze direction, facial expression, body posture, and vocalizations. Every aspect of its design is directed toward making the robot proficient at interpreting and sending readable social cues to the human caregiver, as well as employing a variety of social skills, to foster its behavioral and communication performance (and ultimately its learning performance). This requires that the robot have a rich enough perceptual repertoire to interpret these interactions, and a rich enough behavioral repertoire to act upon them. As such, the design must address the following issues:

Social environment   Kismet must be situated in a social and benevolent learning environment that provides scaffolding interactions. In other words, the environment must contain a benevolent human caregiver.

Real-time performance   Fundamentally, Kismet's world is a social world containing a keenly interesting stimulus: an interested human (sometimes more than one) who is actively trying to engage the robot in a dynamic social manner—to play with it and to teach it about its world. I have found that such a dynamic, complex environment demands a relatively broad and well-integrated perceptual system. For the desired nature and quality of interaction, this system must run at natural interactive rates—in other words, in real-time. The same holds true for the robot's behavioral repertoire and expressive abilities.

Establishment of appropriate social expectations   Kismet should have an appealing appearance and a natural interface that encourages humans to interact with Kismet as if it were a young, socially aware creature. If successful, humans will naturally and unconsciously provide scaffolding interactions. Furthermore, they will expect the robot to behave at a competency-level of an infant-like creature. This level should be commensurate with the robot's perceptual, mechanical, and computational limitations.

Self-motivated interaction   Kismet's synthetic nervous system must motivate the robot to proactively engage in social exchanges with the caregiver and to take an interest in things in the environment. Each social exchange can be viewed as an episode where the robot tries to manipulate the caregiver into addressing its "needs" and "wants." This serves as the basic impetus for social interaction, upon which richer forms of communication can be built. This internal motivation frees the robot from being a slave to its environment, responding only in a reflexive manner to incoming stimuli. Given its own motivations, the robot can internally influence the kinds of interactions it pursues.

Regulation of interactions   Kismet must be capable of regulating the complexity of its interactions with the world and its caregiver. To do this, Kismet should provide the caregiver with social cues (through facial expressions, body posture, or voice) as to whether the interaction is appropriate—i.e., the robot should communicate whether the interaction is overwhelming or under-stimulating. For instance, Kismet should signal to the caregiver when the interaction is overtaxing its perceptual or motor abilities. Further, it should provide readable cues as to what the appropriate level of interaction is. Kismet should exhibit interest in its surroundings and in the humans that engage it, and behave in a way to bring itself closer to desirable aspects and to shield itself from undesirable aspects. By doing so, the robot behaves to promote an environment for which its capabilities are well-matched—ideally, an environment where it is slightly challenged but largely competent—in order to foster its social development.

Readable social cues   Kismet should send social signals to the human caregiver that provide the human with feedback of its internal state. Humans should intuitively and naturally use this feedback to tune their performance in the exchange. Through a process of entraining to the robot, both the human and robot benefit: The person enjoys the easy interaction while the robot is able to perform effectively within its perceptual, computational, and behavioral limits. Ultimately, these cues will allow humans to improve the quality of their instruction.

Interpretation of human's social cues   During social exchanges, the person sends social cues to Kismet to shape its behavior. Kismet must be able to perceive and respond to these cues appropriately. By doing so, the quality of the interaction improves. Furthermore, many of these social cues will eventually be offered in the context of teaching the robot. To be able to take advantage of this scaffolding, the robot must be able to correctly interpret and react to these social cues.

Competent behavior in a complex world   Any convincing robotic creature must address similar behavioral issues as living, breathing creatures. The robot must exhibit robust, flexible, and appropriate behavior in a complex dynamic environment to maintain its "well-being." This often entails having the robot apply its limited resources (finite number of sensors, actuators and limbs, energy, etc.) to perform various tasks. Given a specific task, the robot should exhibit a reasonable amount of persistence. It should work to accomplish a goal, but not at the risk of ignoring other important tasks if the current task is taking too long. Frequently the robot must address multiple goals at the same time. Sometimes these goals are not at cross-purposes and can be satisfied concurrently. Sometimes these goals conflict, and the robot must figure out how to allocate its resources to address both adequately. Which goals the robot pursues, and how it does so, depends both on external influences (from the environment) as well as on internal influences (from the creature's motivations, perceptions, and so forth).

Believable behavior   Operating well in a complex dynamic environment, however, does not ensure convincing, life-like behavior. For Kismet, it is critical that the caregiver perceive the robot as an intentional creature that responds in meaningful ways to her attempts at communication. As previously discussed in chapter 3, the scaffolding the human provides through these interactions is based upon this assumption. Hence, the SNS must address a variety of issues to promote the illusion of a socially aware robotic creature. Blumberg (1996) provides such a list, slightly modified as shown here: *convey intentionality, promote empathy, be expressive,* and *allow variability.*

These are the high-level design issues of the overall human-robot system. The system encompasses the robot, its environment, the human, and the nature of interactions between them. The human brings a complex set of well-established social machinery to the interaction. My aim is not a matter of re-engineering the human side of the equation. Instead, I want to engineer *for* the human side of the equation—to design Kismet's synthetic nervous system to support what comes naturally to people.

If Kismet is designed in a clever manner, people will intuitively engage in appropriate interactions with the robot. This can be accomplished in a variety of ways, such as physically designing the robot to establish the correct set of social expectations for humans, or having Kismet send social cues to humans that they intuitively use to fine-tune their performance.

The following sections present a high-level overview of the SNS. It encompasses the robot's perceptual, motor, attention, motivation, and behavior systems. Eventually, it should include learning mechanisms so that the robot becomes better adapted to its environment over time.

4.2   Design Hints from Animals, Humans, and Infants

In this section, I briefly present ideas for how natural systems address similar issues as those outlined above. Many of these ideas have shaped the design of Kismet's synthetic nervous system. Accordingly, I motivate the high-level design of each SNS subsystem, how each subsystem interfaces with the others, and the responsibility of each for the overall SNS. The following chapters of this book present each subsystem in more detail.

The design of the underlying architecture of the SNS is heavily inspired by models, mechanisms, and theories from the scientific study of intelligent behavior in living creatures. For many years, these fields have sought explanatory models for how natural systems address the aforementioned issues. It is important, however, to distinguish the psychological theory/hypothesis from its underlying implementation in Kismet.

The particular models used to design Kismet's SNS are not necessarily the most recent nor popular in their respective fields. They were chosen based on how easily they could be applied to this application, how compatible they are with other aspects of the system, and how well they could address the relevant issues within synthetic creatures. My focus has been to engineer a system that exhibits the desired behavior, and scientific findings from the study of natural systems have been useful in this endeavor. My aim has not been to explicitly test or verify the validity of these models or theories. Limitations of Kismet's performance could be ascribed to limitations in the mechanics of the implementation (dynamic response of the actuators, processing power, latencies in communication), as well as to the limitations of the models used.

I do not claim explanatory power for understanding human behavior with this implementation. I do not claim equivalence with psychological aspects of human behavior such as emotions, attention, affect, motivation, etc. However, I have implemented synthetic analogs of proposed models, I have integrated them within the same robot, and I have situated Kismet in a social environment. The emergent behavior between Kismet's SNS and its social environment is quite compelling. When I evaluate Kismet, I do so with an engineer's eye. I am testing the adequacy of Kismet's performance, not that of the underlying psychological models.

Below, I highlight special considerations from natural systems that have inspired the design of the robot's SNS. Infants do not come into this world as mindless, flailing skin bags. Instead, they are born as a coherent system, albeit immature, with the ability to respond to and act within their environment in a manner that promotes their survival and continued growth. It is the designer's challenge to bestow upon the robot the innate endowments (i.e., the initial set of software and hardware) that implement similar abilities to that of a newborn. This forms the foundation upon which learning can take place.

Models from ethology have a strong influence in addressing the behavioral issues of the system (e.g., relevance, coherence, concurrency, persistence, and opportunism). As such, they have shaped the manner in which behaviors are organized, expressed, and arbitrated among. Ethology also provides important insights as to how other systems influence behavior (i.e., motivation, perception, attention, and motor expression).

These ethology-based models of behavior are supplemented with models, theories, and behavioral observations from developmental psychology and evolutionary perspectives. In particular, these ideas have had a strong influence in the specification of the "innate endowments" of the SNS, such as early perceptual skills (visual and auditory) and proto-social responses. The field has also provided many insights into the nature of social interaction and learning with a caregiver, and the importance of motivations and emotional responses for this process.

Finally, models from psychology have influenced the design details of several systems. In particular, psychological models of the attention system, facial expressions, the emotion system, and various perceptual abilities have been adapted for Kismet's SNS.

4.3 A Framework for the Synthetic Nervous System

The design details of each system and how they have incorporated concepts from these scientific perspectives are presented in depth in later chapters. Here, I simply present a bird's eye view of the overall synthetic nervous system to give the reader a sense of how the global system fits together. The overall architecture is shown in figure 4.1.

The system architecture consists of six subsystems. The low-level feature extraction system extracts sensor-based features from the world, and the high-level perceptual system encapsulates these features into percepts that can influence behavior, motivation, and motor processes. The attention system determines what the most salient stimulus of the environment is at any time so that the robot can organize its behavior around it. The motivation system regulates and maintains the robot's state of "well-being" in the form of homeostatic regulation processes and emotive responses. The behavior system implements and arbitrates between competing behaviors. The winning behavior defines the current task (i.e., the goal) of the robot. The robot has many behaviors in its repertoire, and several motivations to satiate, so its goals vary over time. The motor system carries out these goals by orchestrating the output modalities (actuator or vocal) to achieve them. For Kismet, these actions are realized as motor skills that accomplish the task physically, or as expressive motor acts that accomplish the task via social signals.

Learning mechanisms will eventually be incorporated into this framework. Most likely, they will be distributed through out the SNS to foster change within various subsystems as well as between them. It is known that natural systems possess many different kinds of interacting learning mechanisms (Gallistel, 1990). Such will be the case with the SNS concerning

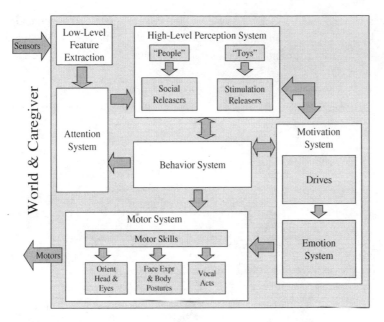

Figure 4.1
A framework for designing synthetic nervous systems. Six sub-systems interact to enable the robot to behave coherently and effectively.

future work. Below, we summarize the systems that comprise the current synthetic nervous system. These can be conceptualized as Kismet's "innate endowments."

The low-level feature extraction system   The low-level feature extraction system is responsible for processing the raw sensory information into quantities that have behavioral significance for the robot. The routines are designed to be cheap, fast, and just adequate. Of particular interest are those perceptual cues that infants seem to rely on. For instance, visual and auditory cues such as detecting eyes and the recognition of vocal affect are important for infants. The low-level perceptual features incorporated into this system are presented in chapters 5, 6, and 7.

The attention system   The low-level visual percepts are sent to the attention system. The purpose of the attention system is to pick out low-level perceptual stimuli that are particularly salient or relevant at that time, and to direct the robot's attention and gaze toward them. This provides the robot with a locus of attention that it can use to organize its behavior. A perceptual stimulus may be salient for several reasons. It may capture the robot's attention because of its sudden appearance, or perhaps due to its sudden change. It may stand out because of its inherent saliency, such as a red ball may stand out from the background. Or perhaps its quality has special behavioral significance for the robot, such as being a typical indication of danger. See chapter 6 and the third CD-ROM demonstration titled "Directing Kismet's Attention" for more details.

The perceptual system   The low-level features corresponding to the target stimuli of the attention system are fed into the perceptual system. Here they are encapsulated into behaviorally relevant percepts. To environmentally elicit processes in these systems, each behavior and emotive response has an associated *releaser*. As conceptualized by Tinbergen (1951) and Lorenz (1973), a releaser can be viewed as a collection of feature detectors that are minimally necessary to identify a particular object or event of behavioral significance. The releasers' function is to ascertain if all environmental (perceptual) conditions are right for the response to become active. High-level perceptions that influence emotive responses are presented in chapter 8, and those that influence task-based behavior are presented in chapter 9.

The motivation system   The motivation system consists of the robot's basic "drives" and "emotions" (see chapter 8). The "drives" represent the basic "needs" of the robot and are modeled as simple homeostatic regulation mechanisms (Carver & Scheier, 1998). When the needs of the robot are being adequately met, the intensity level of each drive is within a desired regime. As the intensity level moves farther away from the homeostatic regime, the robot becomes more strongly motivated to engage in behaviors that restore that drive. Hence, the drives largely establish the robot's own agenda and play a significant role in determining which behavior(s) the robot activates at any one time.

The "emotions" are modeled from a functional perspective. Based on simple appraisals of a given stimulus, the robot evokes either positive emotive responses that serve to bring itself closer to it, or negative emotive responses in order to withdraw from it (refer to the seventh CD-ROM demonstration titled "Emotive Responses"). There is a distinct emotive response for each class of eliciting conditions. Currently, six basic emotive responses are modeled that give the robot synthetic analogs of anger, disgust, fear, joy, sorrow, and surprise (Ekman, 1992). There are also arousal-based responses that correspond to interest, calm, and boredom that are modeled in a similar way. The expression of emotive responses promotes empathy from the caregiver and plays an important role in regulating social interaction with the human. (These expressions are viewable via the second CD-ROM demonstration titled "Readable Expressions.")

The behavior system   The behavior system organizes the robot's task-based behaviors into a coherent structure. Each behavior is viewed as a self-interested, goal-directed entity that competes with other behaviors to establish the current task. An arbitration mechanism is required to determine which behavior(s) to activate and for how long, given that the robot has several motivations that it must tend to and different behaviors that it can use to achieve them. The main responsibility of the behavior system is to carry out this arbitration. In particular, it addresses the issues of relevancy, coherency, persistence, and opportunism. By doing so, the robot is able to behave in a sensible manner in a complex and dynamic environment. The behavior system is described in depth in chapter 9.

The motor system   The motor system arbitrates the robot's motor skills and expressions. It consists of four subsystems: the motor skills system, the facial animation system, the expressive vocalization system, and the oculo-motor system. Given that a particular goal and behavioral strategy have been selected, the motor system determines how to move the robot to carry out that course of action. Overall, the motor skills system coordinates body posture, gaze direction, vocalizations, and facial expressions to address issues of blending and sequencing the action primitives from the specialized motor systems. The motor systems are described in chapters 9, 10, 11, and 12.

4.4   Mechanics of the Synthetic Nervous System

The overall architecture is agent-based as conceptualized by Minsky (1988), Maes (1991), and Brooks (1986), and bears strongest resemblance to that of Blumberg (1996). As such, the SNS is implemented as a highly distributed network of interacting elements. Each computational element (or node) receives messages from those elements connected to its inputs, performs some sort of specific computation based on these messages, and then sends the results to those elements connected to its outputs. The elements connect to form networks, and networks are connected to form the component systems of the SNS.

The basic computational unit   For this implementation, the basic computational process is modeled as shown in figure 4.2. Its *activation level, A*, is computed by the equation: $A = (\sum_{n}^{j=1} w_j \cdot i_j) + b$ for integer values of inputs i_j, weights w_j, and bias b over the number of inputs n. The weights can be either positive or negative; a positive weight corresponds to an excitatory connection, and a negative weight corresponds to an inhibitory connection. Each process is responsible for computing its own activation level. The process is active when its activation level exceeds an *activation threshold, T*. When active, the process can send activation energy to other nodes to favor their activation. It may also perform some special computation, send output messages to connected processes, and/or express itself through motor acts by sending outputs to actuators. Each drive, emotion, behavior, perceptual releaser, and motor process is modeled as a different type that is specifically tailored for its role in the overall system architecture. Hence, although they differ in function, they all follow the basic activation scheme.

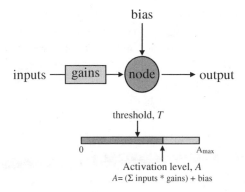

Figure 4.2
A schematic of a basic computational process. The process is active when the activation level A exceeds threshold T.

Networks of units   Units are connected to form networks of interacting processes that allow for more complex computation. This involves connecting the output(s) of one unit to the input(s) of other unit(s). When a unit is active, besides passing messages to the units connected to it, it can also pass some of its activation energy. This is called *spreading activation* and is a mechanism by which units can influence the activation or suppression of other units (Maes, 1991). This mechanism was originally conceptualized by Lorenz (1973) in his *hydraulic model*. Minsky (1988) uses a similar scheme in his ideas of memory formation using *K-lines*.

Subsystems of networks   Groups of connected networks form subsystems. Within each subsystem the active nodes perform special computations to carry out tasks for that subsystem. To do this, the messages that are passed among and within these networks must share a common currency. Thus, the information contained in the messages can be processed and combined in a principled manner (McFarland & Bosser, 1993). Furthermore, as the subsystem becomes more complex, it is possible that some agents may conflict with others (such as when competing for shared resources). In this case, the agents must have some means for competing for expression.

Common currency   This raises an important issue with respect to communication within and between different subsystems. Observable behavior is a product of many interacting processes. Ethology, comparative psychology, and neuroscience have shown that observable behavior is influenced by internal factors (motivations, past experience, etc.) as well as by external factors (perception). This demands that the subsystems be able to communicate and influence each other despite their different functions and modes of computation. This has led ethologists such as McFarland and Bosser (1993) and Lorenz (1973) to propose that there must be a common currency, shared by perceptual, motivational, and behavioral subsystems. In this scheme, the perceptual subsystem generates values based on environmental stimuli, and the motivational subsystem generates values based on internal factors. Both sets of values are passed to the behavioral subsystem, where competing behaviors compute their relevance, based on the perceptual and motivations subsystem values. The two subsystems then compete for expression based on this newly computed value (the common currency).

Within different subsystems, each can operate on their own currencies. This is the case of Kismet's emotion system (chapter 8) and behavior system (chapter 9). The currency that is passed between different systems must be shared, however.

Value-based system   Based upon the use of common currency, the robot's SNS is implemented as a *value-based system*. This simply means that each process computes numeric values (in a common currency) from its inputs. These values are passed as messages (or activation energy) throughout the network, either within a subsystem or between subsystems. Conceptually, the magnitude of the value represents the strength of the contribution in influencing other processes. Using a value-based approach has the nice effect of allowing influences to be graded in intensity, instead of simply being on or off. Other processes compute their relevance based on the incoming activation energies or messages, and use their computed activation level to compete with others for exerting influence upon the SNS.

4.5   Criteria for Evaluation

Thus far in this chapter, I have presented the key design issues for Kismet. To address them, I have outlined the framework for the synthetic nervous system. I now turn to the question of evaluation criteria.

Kismet is neither designed to be a tool nor an interface. One does not use Kismet to perform a task. Kismet is designed to be a robotic creature that can interact socially with humans and ultimately learn from them. As a result, it is difficult or inappropriate to apply standard HCI evaluation criteria to Kismet. Many of these relate to the ability for the system to use natural language, which Kismet is not designed to handle. Some evaluation criteria for embodied conversation agents are somewhat related, such as the use of embodied social cues to regulate turn-taking during dialogues, yet many of these are also closely related to conversational discourse (Sanders & Scholtz, 2000). Currently, Kismet only babbles; it does not speak any natural language.

Instead, Kismet's interactions with humans are fundamentally physical, affective, and social. The robot is designed to elicit interactions with the caregiver that afford rich learning potential. My colleagues and I have endowed the robot with a substantial amount of infrastructure that we believe will enable the robot to leverage from these interactions to foster its social development. As a result, I evaluate Kismet with respect to *interact-ability* criteria. These are inherently subjective, yet quantifiable, measures that evaluate the quality and ease of interaction between robot and human. They address the behavior of both partners, not just the performance of the robot. The evaluation criteria for interact-ability are as follows:

· Do people intuitively read and naturally respond to Kismet's social cues?

· Can Kismet perceive and appropriately respond to these naturally offered cues?

· Does the human adapt to the robot, and the robot adapt to the human, in a way that benefits the interaction? Specifically, is the resulting interaction natural, intuitive, and enjoyable for the human, and can Kismet perform well despite its perceptual, mechanical, behavioral, and computational limitations?

· Does Kismet readily elicit scaffolding interactions from the human that could be used to benefit learning?

4.6 Summary

In this chapter, I have outlined my approach for the design of a robot that can engage humans in a natural, intuitive, social manner. I have carefully considered a set of design issues that are of particular importance when interacting with people (Breazeal, 2001b). Humans will perceive and interpret the robot's actions as socially significant and possessing communicative value. They will respond to them accordingly. This defines a very different set of constraints and challenges for autonomous robot control that lie along a social dimension.

I am interested in giving Kismet the ability to enter into social interactions reminiscent of those that occur between infant and caregiver. These include interactive games, having the human treat Kismet's babbles and expressions as though they are meaningful, and to treat Kismet as a socially aware creature whose behavior is governed by perceived mental states such as intents, beliefs, desires, and feelings. As discussed in chapter 3, these interactions are critical for the social development of infants. Continuing with the infant-caregiver metaphor for Kismet, these interactions could also prove important for Kismet's social development. In chapter 2, I outlined several interesting ways in which various forms of scaffolding address several key challenges of robot learning.

As such, this work is concerned with providing the infrastructure to elicit and support these future learning scenarios. In this chapter, I outlined a framework for this infrastructure that adapts theories, concepts, and models from psychology, social development, ethology, and evolutionary perspectives. The result is a synthetic nervous system that is responsible for generating the observable behavior of the robot and for regulating the robot's internal state of "well-being." To evaluate the performance of both the robot and the human, I introduced a set of evaluation criteria for interact-ability. Throughout the book, I will present a set of studies with naive human subjects that provide the data for our evaluations. In the following chapter, I begin my in-depth presentation of Kismet starting with a description of the physical robot and its computational platform.

5 The Physical Robot

The design task is to build a physical robot that encourages humans to treat it as if it were a young socially aware creature. The robot should therefore have an appealing infant-like appearance so that humans naturally fall into this mode of interaction. The robot must have a natural and intuitive interface (with respect to its inputs and outputs) so that a human can interact with it using natural communication channels. This enables the robot to both read and send human-like social cues. Finally, the robot must have sufficient sensory, motor, and computational resources for real-time performance during dynamic social interactions with people.

5.1 Robot Aesthetics and Physicality

When designing robots that interact socially with people, the aesthetics of the robot should be carefully considered. The robot's physical appearance, its manner of movement, and its manner of expression convey personality traits to the person who interacts with it. This fundamentally influences the manner in which people engage the robot.

Youthful and appealing It will be quite a while before we are able to build autonomous humanoids that rival the social competence of human adults. For this reason, Kismet is designed to have an infant-like appearance of a fanciful robotic creature. Note that the human is a critical part of the environment, so evoking appropriate behaviors from the human is essential for this project. The key set of features that evoke nurturing responses of human adults (see figure 5.1) has been studied across many different cultures (Eibl-Eibesfeldt, 1972), and these features have been explicitly incorporated into Kismet's design (Breazeal & Foerst, 1999). Other issues such as physical size and stature also matter. For instance, when people are standing they look down to Kismet and when they are seated they can engage the robot at eye level. As a result, people tend to intuitively treat Kismet as a very young creature and modify their behavior in characteristic baby-directed ways. As argued in chapter 3, the same characteristics could be used to benefit the robot by simplifying the perceptual challenges it faces when behaving in the physical world. It also allows the robot to participate in interesting social interactions that are well-matched to the robot's level of competence.

Believable versus realistic Along a similar vein, the design should minimize factors that could detract from a natural infant-caretaker interaction. Ironically, humans are particularly sensitive (in a negative way) to systems that try to imitate humans but inevitably fall short. Humans have strong implicit assumptions regarding the nature of human-like interactions, and they are disturbed when interacting with a system that violates these assumptions (Cole, 1998). For this reason, I consciously decided to *not* make the robot look human. Instead

Figure 5.1
Examples of the baby scheme of Eibl-Eibesfeldt (1972). He posits that a set of facial characteristics cross-culturally trigger nurturing responses from adults. These include a large head with respect to the body, large eyes with respect to the face, a high forehead, and lips that suggest the ability to suck. These features are commonly incorporated into dolls and cartoons, as shown here.

the robot resembles a young, fanciful creature with anthropomorphic expressions that are easily recognizable to a human.

As long argued by animators, a character does not have to be realistic to be believable—i.e., to convey the illusion of life and to portray a thinking and feeling being (Thomas & Johnston, 1981). Ideally, people will treat Kismet as if it were a socially aware creature with thoughts, intents, desires, and feelings. Believability is the goal. Realism is not necessary.

Audience perception   A deep appreciation of audience perception is a fundamental issue for classical animation (Thomas & Johnston, 1981) and has more recently been argued for by Bates (1994) in his work on believable agents. For sociable robots, this issue holds as well (albeit for different reasons) and can be experienced firsthand with Kismet. How the human perceives the robot establishes a set of expectations that fundamentally shape how the human interacts with it. This is not surprising as Reeves and Nass (1996) have demonstrated this phenomenon for media characters, cartoon characters, as well as embodied conversation agents.

Being aware of these social factors can be played to advantage by establishing an appropriate set of expectations through robotic design. If done properly, people tend to naturally tune their behavior to the robot's current level of competence. This leads to a better quality of interaction for both robot and human.

5.2   The Hardware Design

Kismet is an expressive robotic creature with perceptual and motor modalities tailored to natural human communication channels. To facilitate a natural infant-caretaker interaction, the robot is equipped with input and output modalities roughly analogous to those of an infant (of course, missing many that infants have). For Kismet, the inputs include visual, auditory, and proprioceptive sensory inputs.

The motor outputs include vocalizations, facial expressions, and motor capabilities to adjust the gaze direction of the eyes and the orientation of the head. Note that these motor systems serve to steer the visual and auditory sensors to the source of the stimulus and can also be used to display communicative cues. The choice of these input and output modalities is geared to enable the system to participate in social interactions with a human, as opposed to traditional robot tasks such as manipulating physical objects or navigating through a cluttered space. Kismet's configuration is most clearly illustrated by watching the included CD-ROM's introductory "What is Kismet?" section. A schematic of the computational hardware is shown in figure 5.2.

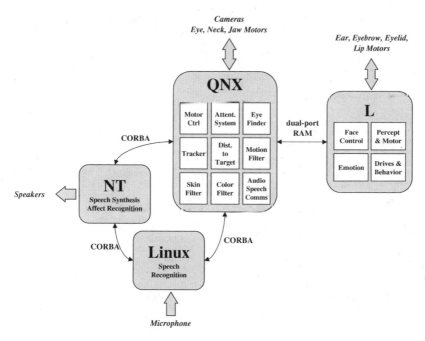

Figure 5.2
Kismet's hardware and software control architectures have been designed to meet the challenge of real-time processing of visual signals (approaching 30 Hz) and auditory signals (8 kHz sample rate and frame windows of 10 ms) with minimal latencies (less than 500 ms). The high-level perception system, the motivation system, the behavior system, the motor skills system, and the face motor system execute on four Motorola 68332 microprocessors running L, a multi-threaded Lisp developed in our lab. Vision processing, visual attention, and eye/neck control are performed by nine networked 400 MHz PCs running QNX (a real-time Unix-like operating system). Expressive speech synthesis and vocal affective intent recognition runs on a dual 450 MHz PC running Windows NT, and the speech recognition system runs on a 500 MHz PC running Linux.

The Vision System

The robot's vision system consists of four color CCD cameras mounted on a stereo active vision head. Two wide field of view (FoV) cameras are mounted centrally and move with respect to the head. These are 0.25 inch CCD lipstick cameras with 2.2 mm lenses manufactured by Elmo Corporation. They are used to direct the robot's attention toward people or toys and to compute a distance estimate. There is also a camera mounted within the pupil of each eye. These are 0.5 inch CCD foveal cameras with an 8 mm focal length lenses, and are used for higher resolution post-attentional processing, such as eye detection.

Kismet has three degrees of freedom to control gaze direction and three degrees of freedom (DoF) to control its neck (see figure 5.3). Each eye has an independent pan DoF, and both eyes share a common tilt DoF. The degrees of freedom are driven by Maxon DC servo motors with high resolution optical encoders for accurate position control. This gives the robot the ability to move and orient its eyes like a human, engaging in a variety of human visual behaviors. This is not only advantageous from a visual processing perspective (as advocated by the active vision community such as Ballard [1989]), but humans attribute a communicative value to these eye movements as well. For instance, humans use gaze direction to infer whether a person is attending to them, to an object of shared interest, or neither. This is important information when trying to carry out face-to-face interaction.

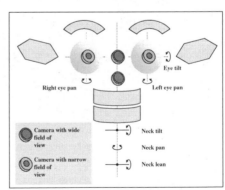

Figure 5.3
Kismet has a large set of expressive features—eyelids, eyebrows, ears, jaw, lips, neck, and eye orientation. The schematic on the right shows the degrees of freedom (DoF) relevant to visual perception (omitting the eyelids). The eyes can turn independently along the horizontal (pan), but only turn together along the vertical (tilt). The neck can turn the whole head horizontally and vertically, and can also lean forward or backward. Two cameras with narrow "foveal" fields of view rotate with the eyes. Two central cameras with wide fields of view rotate with the neck. These cameras are unaffected by the orientation of the eyes. Please refer to the CD-ROM section titled "What is Kismet?"

Kismet's vision system is implemented on a network of nine 400 MHz commercial PCs running the QNX real-time operating system. The PCs are connected together via 100 MB Ethernet. There are frame grabbers and video distribution amplifiers to distribute multiple copies of a given image with minimal latencies. The cameras that are used to compute stereo measures are externally synchronized.

The Auditory System

The caregiver can influence the robot's behavior through speech by wearing a small unobtrusive wireless microphone. This auditory signal is fed into a 500 MHz PC running Linux. The real-time, low-level speech processing and recognition software was developed at MIT by the Spoken Language Systems Group. These auditory features are sent to a dual 450 mHz PC running Windows NT. The NT machine processes these features in real-time to recognize the spoken affective intent of the caregiver. The Linux and NT machines are connected via 100 MB Ethernet to a shared hub and use CORBA for communication.

The Expressive Motor System

Kismet is able to display a wide assortment of facial expressions that mirror its affective state, as well as produce numerous facial displays for other communicative purposes (Breazeal & Scassellati, 1999b). Figure 5.4 illustrates a few examples. All eight expressions, and their accompanying vocalizations, are shown in the second demonstration on the included CD-ROM. Fourteen of the face actuators are Futaba micro servos, which come in a lightweight and compact package. Each ear has two degrees of freedom that enable each to elevate and rotate. This allows the robot to perk its ears in an interested fashion, or fold them back in a manner reminiscent of an angry animal. Each eyebrow has two degrees of freedom that enable each to elevate and to arc toward and away from the centerline. This allows the brows to furrow in frustration, or to jolt upward in surprise. Each eyelid can open and close independently, allowing the robot to wink an eye or blink both. The robot has four

Figure 5.4
Some example facial expressions that illustrate the movement of Kismet's facial features. From left to right they correspond to expressions for sadness, disapproval, happiness, and surprise.

lip actuators, two for the upper lip corners and two for the lower lip corners. Each actuator moves a lip corner either up (to smile), or down (to frown). There is also a single degree of freedom jaw that is driven by a high performance DC servo motor from the MEI card. This level of performance is important for real-time lip synchronization with speech.

The face control software runs on a Motorola 68332 node running L. This processor is responsible for arbitrating between facial expression, real-time lip synchronization, communicative social displays, as well as behavioral responses. It communicates to other 68332 nodes through a 16 KByte dual-ported RAM (DPRAM).

High-Level Perception, Behavior, Motivation, and Motor Skills

The high-level perception system, the behavior system, the motivation system, and the motor skills system run on the network of Motorola 68332 micro-controllers. Each of these systems communicates with the others by using threads if they are implemented on the same processor, or via DPRAM communication if implemented on different processors. Currently, each 68332 node can hook up to at most eight DPRAMs. Another single DPRAM tethers the 68332 network to the network of PC machines via a QNX node.

The Vocalization System

The robot's vocalization capabilities are generated through an articulatory synthesizer. The software, *DECtalk v4.5* sold by Digital Equipment Corporation, is based on the Klatt articulation synthesizer and runs on a PC under Windows NT with a Creative Labs sound card. The parameters of the model are based on the physiological characteristics of the human articulatory tract. Although typically used as a text-to-speech system, it was chosen over other systems because it gives the user low-level control over the vocalizations through physiologically based parameter settings. These parameters make it possible to convey affective information through vocalizations (Cahn, 1990), and to convey personality by designing a custom voice for the robot. As such, Kismet's voice is that of a young child. The system also has the ability to play back files in a `.wav` format, so the robot could in principle produce infant-like vocalizations (laughter, coos, gurgles, etc.) that the synthesizer itself cannot generate.

Instead of relying on written text as an interface to the synthesizer, the software can accept strings of phonemes along with commands to specify the pitch and timing of the utterance. Hence, Kismet's vocalization system generates both phoneme strings and command settings, and says them in near real-time. The synthesizer also extracts phoneme and pitch information that are used to coordinate real-time lip synchronization. Ultimately, this capability would permit the robot to play and experiment with its own vocal tract, and to learn the effect these vocalizations have on human behavior. Kismet's voice is one of the most versatile instruments it has to interact with the caregiver. Examples of these vocalizations can be heard by watching the "Readable Expressions" demonstration on the included CD-ROM.

5.3   Overview of the Perceptual System

Human infants discriminate readily between social stimuli (faces, voices, etc.) and salient non-social stimuli (brightly colored objects, loud noises, large motion, etc.). For Kismet, the perceptual system is designed to discriminate a subset of both social and non-social stimuli from visual images as well as auditory streams. The specific percepts within each category (social versus non-social) are targeted for social exchanges. Specifically, the social stimuli are geared toward detecting the affective state of the caregiver, whether or not the caregiver is paying attention to the robot, and other people-related percepts that are important during face-to-face exchanges such as the prosody of the caregiver's vocalizations. The non-social percepts are selected for their ability to command the attention of the robot. These are useful during social exchanges when the caregiver wants to direct the robot's attention to events outside pure face-to-face exchange. In this way, the caregiver can focus the interaction on things and events in the world, such as centering an interaction around playing with a specific toy.

Our discussion of the perceptual limitations of infants in chapter 3 has important implications for how to design Kismet's perceptual system. Clearly the ultimate, most versatile and complete perceptual system is not necessary. A perceptual system that rivals the performance and sophistication of the adult is not necessary either. As argued in chapter 3, this is not appropriate and would actually hinder development by overwhelming the robot with more perceptual information than the robot's synthetic nervous system could possibly handle or learn from. It is also inappropriate to place the robot in an overly simplified environment where it would ultimately learn and predict everything about that environment. There would be no impetus for continued growth. Instead, the perceptual system should start out as simple as possible, but rich enough to distinguish important social cues and interaction scenarios that are typical of caregiver-infant interactions. In the meantime, the caregiver must do her part to simplify the robot's perceptual task by slowing down and exaggerating her behavior in appropriate ways. She should repeat her behavior until she feels it has been adequately perceived by the robot, so the robot does not need to get the perception exactly right upon its first appearance. The challenge is to specify a perceptual system that can detect the right kinds of information at the right resolution.

A relatively broad and well-integrated real-time perceptual system is critical for Kismet's success in the infant-caregiver scenario. The real-time constraint imposes some fairly stringent restrictions in the algorithms used. As a result, these algorithms tend to be simple and of low resolution so that they can run quickly. One might characterize Kismet's perceptual system as being broad and simple where the perceptual abilities are robust enough and detailed enough for these early human-robot interactions. Deep and complicated perceptual algorithms certainly exist. As we have learned from human infants, however, there are developmental advantages to starting out broad and simple and allowing the perceptual, behavioral, and motor systems to develop in step. Kismet's initial perceptual system specification is designed to be roughly analogous to a human infant. While human infants certainly perceive more things than Kismet, it is quite a sophisticated perceptual system for an autonomous robot.

The perceptual system is decomposed into six subsystems (see figure 5.5). The development of Kismet's overall perceptual system is a large-scale engineering endeavor that includes the efforts of many collaborators. I include citations wherever possible, although some work has yet to be published. Please see the preface where I gratefully recognize the efforts of these researchers. I describe the visual attention system in chapter 6. I cover the affective speech recognition system in chapter 7. The behavior-specific and emotion-specific perceptions (organized around the social/non-social perceptual categories) are discussed in chapters 8 and 9. For the remainder of this chapter, I briefly outline the low-level perceptual abilities for visual and auditory channels.

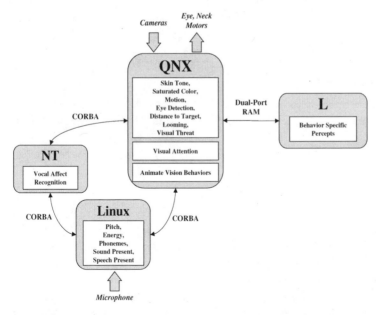

Figure 5.5
Schematic of Kismet's perceptual systems.

Low-Level Visual Perception

Kismet's low-level visual perception system extracts a number of features that human infants seem to be particularly responsive toward. These low-level features were selected for their ability to help Kismet distinguish social stimuli (i.e., people, based on skin tone, eye detection, and motion) from non-social stimuli (i.e., toys, based on saturated color and motion), and to interact with each in interesting ways (often modulated by the distance of the target stimulus to the robot). There are a few perceptual abilities that serve self-protection responses. These include detecting looming stimuli as well as potentially dangerous stimuli (characterized by excessive motion close to the robot). We have previously reported an overview of Kismet's visual abilities (Breazeal et al., 2000; Breazeal & Scassellati, 1999a,b). Kismet's low-level visual features are as follows (in parentheses, I gratefully acknowledge my colleagues who have implemented these perceptual abilities on Kismet):

- Highly saturated color: red, blue, green, yellow (B. Scassellati)
- Colors representative of skin tone (P. Fitzpatrick)
- Motion detection (B. Scasselatti)
- Eye detection (A. Edsinger)
- Distance to target (P. Fitzpatrick)
- Looming (P. Fitzpatrick)
- Threatening, very close, excessive motion (P. Fitzpatrick)

Low-Level Auditory Perception

Kismet's low-level auditory perception system extracts a number of features that are also useful for distinguishing people from other sound-emitting objects such as rattles and bells. The software runs in real-time and was developed at MIT by the Spoken Language Systems Group (www.sls.lcs.mit.edu/sls). Jim Glass and Lee Hetherington were tremendously helpful in tailoring the code for Kismet's specific needs and in helping port this sophisticated speech recognition system to Kismet. The software delivers a variety of information that is used to distinguish speech-like sounds from non-speech sounds, to recognize vocal affect, and to regulate vocal turn-taking behavior. The phonemic information may ultimately be used to shape the robot's own vocalizations during imitative vocal games, and to enable the robot to acquire a proto-language from long-term interactions with human caregivers. Kismet's low-level auditory features are as follows:

- Sound present
- Speech present

- Time-stamped pitch tracking
- Time-stamped energy tracking
- Time-stamped phonemes

5.4   Summary

Kismet is an expressive robotic creature with perceptual and motor modalities tailored to natural human communication channels. To facilitate a natural infant-caretaker interaction, the robot is equipped with visual, auditory, and proprioceptive sensory inputs. Its motor modalities consist of a high-performance six DoF active vision head supplemented with expressive facial features. Its hardware and software control architectures have been designed to meet the challenge of real-time processing of visual signals (approaching 30 Hz) and auditory signals (frame windows of 10 ms) with minimal latencies (<500 ms). These fifteen networked computers run the robot's synthetic nervous system that integrates perception, attention, motivations, behaviors, and motor acts.

Kismet's perceptual system is designed to support a variety of important functions. Many aspects address behavioral and protective responses that evolution has endowed to living creatures so that they may behave and survive in the physical world. Given the perceptual richness and complexity of the physical world, I have implemented specific systems to explicitly organize this flood of information. By doing so, the robot can organize its behavior around a locus of attention.

The robot's perceptual abilities have been explicitly tailored to support social interaction with people and to support social learning/instruction processes. The robot must share enough of a perceptual world with humans so that communication can take place. The robot must be able to perceive the social cues that people naturally and intuitively use to communicate with it. The robot and a human should share enough commonality in those features of the perceptual world that are of particular interest, so that both are drawn to attend to similar events and stimuli. Meeting these criteria enables a human to naturally and intuitively direct the robot's attention to interesting things in order to establish shared reference. It also allows a human to communicate affective assessments to the robot, which could make social referencing possible. Ultimately these abilities will play an important role in the robot's social development, as they do for the social development of human infants.

6 The Vision System

Certain types of spontaneously occurring events may momentarily dominate his attention or cause him to react in a quasi-reflex manner, but a mere description of the classes of events which dominate and hold the infants' sustained attention quickly leads one to the conclusion that the infant is biologically tuned to react to person-mediated events. These being the only events he is likely to encounter which will be phased, in their timing, to coordinate in a non-predictable or non-redundant way with his own activities and spontaneous reactions.
—J. Newson (1979, p. 207)

There are a number of stimuli that infants have a bias to attend to. They can be categorized according to visual versus auditory sensory channels (among others), and whether they correspond to social forms of stimulation. Accordingly, similar percepts have been implemented on Kismet because of their important role in social interaction. Of course, there are other important features that have yet to be implemented. The attention system (designed in collaboration with Brian Scassellati) directs the robot's attention to those visual sensory stimuli that can be characterized by these selected perceptions. Later extensions to the mechanism could include other perceptual features.

To benefit communication and social learning, it is important that both robot and human find the same sorts of perceptual features interesting. Otherwise there will be a mismatch between the sorts of stimuli and cues that humans use to direct the robot's attention versus those that attract the robot's attention. If designed improperly, it could prove to be very difficult to achieve joint reference with the robot. Even if the human could learn what attracts the robot's attention, this defeats the goal of allowing the person to use natural and intuitive cues. Designing for the set of perceptual cues that human infants find salient allows us to implement an initial set that are naturally significant for humans.

6.1 Design of the Attention System

Kismet's attention system acts to direct computational and behavioral resources toward salient stimuli and to organize subsequent behavior around them. In an environment suitably complex for interesting learning, perceptual processing will invariably result in many potential target stimuli. It is critical that this be accomplished in real-time. In order to determine where to assign resources, the attention system must incorporate raw sensory saliency with task-driven influences.

The attention system is shown in figure 6.1 and is heavily inspired by the *Guided Search v2.0* system of Wolfe (1994). Wolfe proposed this work as a model for human visual search behavior. Brian Scassellati and I have extended it to account for moving cameras, dynamically changing task-driven influences, and habituation effects (Breazeal & Scassellati, 1999a). The accompanying CD-ROM also includes a video demonstration of the attention system as its third demo, "Directing Kismet's Attention."

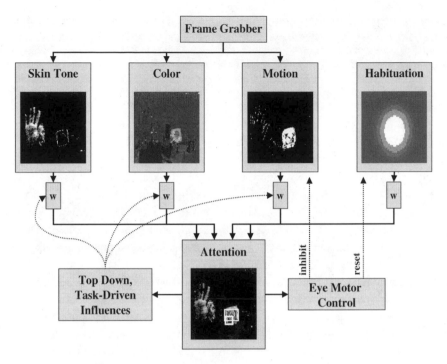

Figure 6.1
The robot's attention is determined by a combination of low-level perceptual stimuli. The relative weightings of the stimuli are modulated by high-level behavior and motivational influences. A sufficiently salient stimulus in any modality can preempt attention, similar to the human response to sudden motion. All else being equal, larger objects are considered more salient than smaller ones. The design is intended to keep the robot responsive to unexpected events, while avoiding making it a slave to every whim of its environment. With this model, people intuitively provide the right cues to direct the robot's attention (shake object, move closer, wave hand, etc.). Displayed images were captured during a behavioral trial session.

The attention system is a two-stage system. The first stage is a *pre-attentive,* massively parallel stage that processes information about basic visual features (e.g., color, motion, depth cues) across the entire visual field (Triesman, 1986). For Kismet, these bottom-up features include highly saturated color, motion, and colors representative of skin tone. The second stage is a *limited capacity* stage that performs other more complex operations, such as facial expression recognition, eye detection, or object identification, over a localized region of the visual field. These limited capacity processes are deployed serially from location to location under attentional control. This is guided by the properties of the visual stimuli processed by the first stage (an exogenous contribution), by task-driven influences, and by habituation effects (both are endogenous contributions). The habituation influence provides Kismet with a primitive attention span. For Kismet, the second stage includes an eye-detector that operates over the foveal image, and a target proximity estimator that operates on the stereo images of the two central wide field of view (FoV) cameras.

Four factors (pre-attentive processing, post-attentive processing, task-driven influences, and habituation) influence the direction of Kismet's gaze. This in turn determines the robot's subsequent perception, which ultimately feeds back to behavior. Hence, the robot is in a continuous cycle: behavior influencing what is perceived, and perception influencing subsequent behavior.

Bottom-up Contributions: Computing Feature Maps

The purpose of the first massively parallel stage is to identify locations that are worthy of further attention. This is considered to be a *bottom-up* or stimulus-driven contribution. Raw sensory saliency cues are equivalent to those "pop-out" effects studied by Triesman (1986), such as color intensity, motion, and orientation for visual stimuli. As such, it serves to bias attention toward distinctive items in the visual field and will not guide attention if the properties of that item are not inherently salient.

This contribution is computed from a series of *feature maps,* which are updated in parallel over the entire visual field (of the wide FoV camera) for a limited set of basic visual features. There is a separate feature map for each basic feature (for Kismet these correspond to color, motion, and skin tone), and each map is topographically organized and in retinotopic coordinates. The computation of these maps is described below. The value of each location is called the *activation level* and represents the saliency of that location in the visual field with respect to the other locations. In this implementation, the overall bottom-up contribution comes from combining the results of these feature maps in a weighted sum.

The video signal from each of Kismet's cameras is digitized by one of the 400 MHz nodes with frame-grabbing hardware. The image is then subsampled and averaged to an appropriate size. Currently, we use an image size of 128×128, which allows us to complete all of the processing in near real-time. To minimize latency, each feature map is computed by a separate 400 MHz processor (each of which also has additional computational task load). All of the feature detectors discussed here can operate at multiple scales.

Color saliency feature map   One of the most basic and widely recognized visual features is color. These models of color saliency are drawn from the complementary work on visual search and attention (Itti et al., 1998). The incoming video stream contains three 8-bit color channels (r for red, g for green, and b for blue) each with a 0 to 255 value range that are transformed into four color-opponent channels (r', g', b', and y'). Each input color channel is first normalized by the luminance l (a weighted average of the three input color channels):

$$r_n = \frac{255}{3} \cdot \frac{r}{l} \qquad g_n = \frac{255}{3} \cdot \frac{g}{l} \qquad b_n = \frac{255}{3} \cdot \frac{b}{l} \tag{6.1}$$

These normalized color channels are then used to produce four opponent-color channels:

$$r' = r_n - (g_n + b_n)/2 \tag{6.2}$$

$$g' = g_n - (r_n + b_n)/2 \tag{6.3}$$

$$b' = b_n - (r_n + g_n)/2 \tag{6.4}$$

$$y' = \frac{r_n + g_n}{2} - b_n - \|r_n - g_n\| \tag{6.5}$$

The four opponent-color channels are clamped to 8-bit values by thresholding. While some research seems to indicate that each color channel should be considered individually (Nothdurft, 1993), Scassellati chose to maintain all of the color information in a single feature map to simplify the processing requirements (as docs Wolfe [1994] for more theoretical reasons). The result is a two-dimensional map where pixels containing a bright, saturated color component (red, green, blue, and yellow) have a greater intensity value. Kismet is particularly sensitive to bright red, green, yellow, blue, and even orange. Figure 6.1 gives an example of the color feature map when the robot looks at a brightly colored block.

Motion saliency feature maps   In parallel with the color saliency computations, a second processor receives input images from the frame grabber and computes temporal differences to detect motion. Motion detection is performed on the wide FoV camera, which is often at rest since it does not move with the eyes. The incoming image is converted to grayscale and placed into a ring of frame buffers. A raw motion map is computed by passing the absolute difference between consecutive images through a threshold function $\mathcal{T}$:

$$M_{raw} = \mathcal{T}(\|I_t - I_{t-1}\|) \tag{6.6}$$

This raw motion map is then smoothed with a uniform 7×8 field. The result is a binary 2-D map where regions corresponding to motion have a high intensity value. The motion saliency feature map is computed at 25-30 Hz by a single 400 MHz processor node. Figure 6.1 gives an example of the motion feature map when the robot looks at a toy block that is being shaken.

Skin tone feature map   Colors consistent with skin are also filtered for. This is a computationally inexpensive means to rule out regions that are unlikely to contain faces or

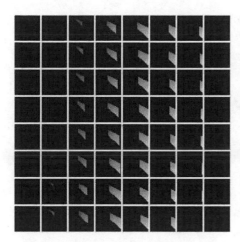 

Figure 6.2
The skin tone filter responds to 4.7 percent of possible (R, G, B) values. Each grid element in the figure to the left shows the response of the filter to all values of red and green for a fixed value of blue. Within a cell, the x-axis corresponds to red and the y-axis corresponds to green. The image to the right shows the filter in operation. Typical indoor objects that may also be consistent with skin tone include wooden doors, pink walls, etc.

hands. Most pixels on faces will pass these tests over a wide range of lighting conditions and skin color. Pixels that pass these tests are weighted according to a function learned from instances of skin tone from images taken by Kismet's cameras (see figure 6.2). In this implementation, a pixel is *not* skin-toned if:

- $r < 1.1 \cdot g$ (the red component fails to dominate green sufficiently)
- $r < 0.9 \cdot b$ (the red component is excessively dominated by blue)
- $r > 2.0 \cdot max(g, b)$ (the red component completely dominates both blue and green)
- $r < 20$ (the red component is too low to give good estimates of ratios)
- $r > 250$ (the red component is too saturated to give a good estimate of ratios)

Top-down Contributions: Task-Based Influences

For a goal-achieving creature, the behavioral state should also bias what the creature attends to next. For instance, when performing visual search, humans seem to be able to preferentially select the output of one broadly tuned channel per feature (e.g., "red" for color and "shallow" for orientation if searching for red horizontal lines) (Kandel et al., 2000).

For Kismet, these top-down, behavior-driven factors modulate the output of the individual feature maps before they are summed to produce the bottom-up contribution. This process selectively enhances or suppresses the contribution of certain features, but does not alter the underlying raw saliency of a stimulus (Niedenthal & Kityama, 1994). To implement this, the bottom-up results of each feature map are each passed through a filter (effectively a gain). The value of each gain is determined by the active behavior. These modulated feature maps are then summed to compute the overall attention activation map.

This serves to bias attention in a way that facilitates achieving the goal of the active behavior. For example, if the robot is searching for social stimuli, it becomes sensitive to skin tone and less sensitive to color. Behaviorally, the robot may encounter toys in its search, but will continue until a skin-toned stimulus is found (often a person's face). Figure 6.3 illustrates how gain adjustment biases what the robot finds to be more salient.

As shown in figure 6.4, the skin-tone gain is enhanced when the `seek-people` behavior is active, and is suppressed when the `avoid-people` behavior is active. Similarly, the color gain is enhanced when the `seek-toys` behavior is active, and suppressed when the `avoid-toys` behavior is active. Whenever the `engage-people` or `engage-toys` behaviors are active, the face and color gains are restored to slightly favor the desired stimulus. Weight adjustments are constrained such that the total sum of the weights remains constant at all times.

Figure 6.3
Effect of gain adjustment on looking preference. Circles correspond to fixation points, sampled at one-second intervals. On the left, the gain of the skin tone filter is higher. The robot spends more time looking at the face in the scene (86% face, 14% block). This bias occurs despite the fact that the face is dwarfed by the block in the visual scene. On the right, the gain of the color saliency filter is higher. The robot now spends more time looking at the brightly colored block (28% face, 72% block).

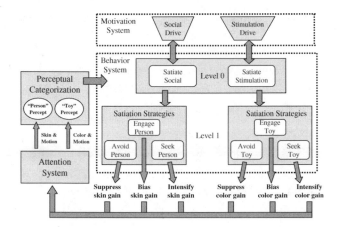

Figure 6.4
Schematic of behaviors relevant to attention. The activation of a particular behavior depends on both perceptual factors and motivation factors. The "drives" within the motivation system have an indirect influence on attention by influencing the behavioral context. The behaviors at Level One of the behavior system directly manipulate the gains of the attention system to benefit their goals. Through behavior arbitration, only one of these behaviors is active at any time.

Computing the Attention Activation Map

The attention activation map can be thought of as an activation "landscape" with higher hills marking locations receiving substantial bottom-up or top-down activation. The purpose of the attention activation map (using the terminology of Wolfe) is to direct attention, where attention is attracted to the highest hill. The greater the activation at a location, the more likely the attention will be directed to that location. Note that by using this approach, the locus of activation contains no information as to its source (i.e., a high activation for color looks the same as high activation for motion information). The activation map makes it possible to guide attention based on information from more than one feature (such as a conjunction of features).

To prevent drawing attention to non-salient regions, the attention activation map is thresholded to remove noise values and normalized by the sum of the gains. Connected object regions are extracted using a grow-and-merge procedure with 4-connectivity (Horn, 1986). To further combine related regions, any regions whose bounding boxes have a significant overlap are also merged. The attention process runs at 20 Hz on a single 400 MHz processor.

Statistics on each region are then collected, including the centroid, bounding box, area, average attention activation score, and average score for each of the feature maps in that region. The tagged regions that are large enough (having an area of at least thirty pixels) are sorted based upon their average attention activation score. The attention process provides the top three regions to both the eye motor control system and the behavior and motivational systems.

The most salient region is the new visual target. The individual feature map scores of the target are passed onto higher-level perceptual stages where these features are combined to form behaviorally meaningful percepts. Hence, the robot's subsequent behavior is organized about this locus of attention.

Attention Drives Eye Movement

Gaze direction is a powerful social cue that people use to determine what interests others. By directing the robot's gaze to the visual target, the person interacting with the robot can accurately use the robot's gaze as an indicator of what the robot is attending to. This greatly facilitates the interpretation and readability of the robot's behavior, since the robot reacts specifically to the thing that it is looking at.

The eye-motor control system uses the centroid of the most salient region as the target of interest. The eye-motor control process acts on the data from the attention process to center the eyes on an object within the visual field. Using a data-driven mapping between image position and eye position, the retinotopic coordinates of the target's centroid are used to compute where to look next (Scassellati, 1998). Each time that the neck moves, the eye/neck motor process sends two signals. The first signal inhibits the motion detection system for approximately 600 ms, which prevents self-motion from appearing in the motion feature map. The second signal resets the habituation state, described in the next section. A detailed discussion of how the motor component from the attention system is integrated into the rest of Kismet's visual behavior (such as smooth pursuit, looming, etc.) appears in chapter 12. Kismet's visual behavior can be seen in the sixth CD-ROM demonstration titled "Visual Behaviors."

Habituation Effects

To build a believable creature, the attention system must also implement habituation effects. Infants respond strongly to novel stimuli, but soon habituate and respond less as familiarity increases (Carey & Gelman, 1991). This acts both to keep the infant from being continually fascinated with any single object and to force the caregiver to continually engage the infant with slightly new and interesting interactions. For a robot, a habituation mechanism removes the effects of highly salient background objects that are not currently involved in direct interactions as well as placing requirements on the caregiver to maintain interaction with different kinds of stimulation.

To implement habituation effects, a *habituation filter* is applied to the activation map over the location currently being attended to. The habituation filter effectively decays the activation level of the location currently being attended to, strengthening bias toward other locations of lesser activation.

The habituation function can be viewed as a feature map that initially maintains eye fixation by increasing the saliency of the center of the field of view and then slowly decays the saliency values of central objects until a salient off-center object causes the neck to move. The habituation function is a Gaussian field $G(x, y)$ centered in the field of view with peak amplitude of 255 (to remain consistent with the other 8-bit values) and $\theta = 50$ pixels. It is combined linearly with the other feature maps using the weight

$$w = W \cdot max(-1, 1 - \Delta t/\tau) \qquad (6.7)$$

where w is the weight, Δt is the time since the last habituation reset, τ is a time constant, and W is the maximum habituation gain. Whenever the neck moves, the habituation function is reset, forcing w to W and amplifying the saliency of central objects until a time τ when $w = 0$ and there is no influence from the habituation map. As time progresses, w decays to a minimum value of $-W$ which suppresses the saliency of central objects. In the current implementation, a value of $W = 10$ and a time constant $\tau = 5$ seconds is used. When the robot's neck shifts, the habituation map is reset, allowing that region to be revisited after some period of time.

6.2 Post-Attentive Processing

Once the attention system has selected regions of the visual field that are potentially behaviorally relevant, more intensive computation can be applied to these regions than could be applied across the whole field. Searching for eyes is one such task. Locating eyes is important to us for engaging in eye contact. Eyes are searched for after the robot directs its gaze to a locus of attention. By doing so, a relatively high-resolution image of the area being searched is available from the narrow FoV cameras (see figure 6.5).

Once the target of interest has been selected, its proximity to the robot is estimated using a stereo match between the two central wide FoV cameras. Proximity is an important factor for interaction. Things closer to the robot should be of greater interest. It is also useful for interaction at a distance. For instance, a person standing too far from Kismet for face-to-face interaction may be close enough to be beckoned closer. Clearly the relevant behavior (beckoning or playing) is dependent on the proximity of the human to the robot.

Eye detection   Detecting people's eyes in a real-time robotic domain is computationally expensive and prone to error due to the large variance in head posture, lighting conditions and feature scales. Aaron Edsinger developed an approach based on successive feature extraction, combined with some inherent domain constraints, to achieve a robust and fast

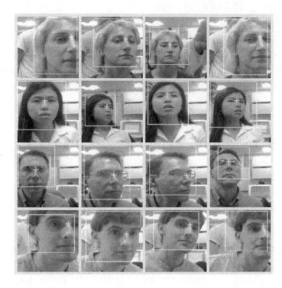

Figure 6.5
Sequence of foveal images with eye detection. The eye detector actually looks for the region between the eyes. The box indicates a possible face has been detected (being both skin-toned and oval in shape). The small cross locates the region between the eyes.

eye-detection system for Kismet (Breazeal et al., 2001). First, a set of feature filters are applied successively to the image in increasing feature granularity. This serves to reduce the computational overhead while maintaining a robust system. The successive filter stages are:

• Detect skin-colored patches in the image (abort if this does not pass above a threshold).

• Scan the image for ovals and characterize its skin tone for a potential face.

• Extract a sub-image of the oval and run a ratio template over it for candidate eye locations (Sinha, 1994; Scassellati, 1998).

• For each candidate eye location, run a pixel-based multi-layer perceptron (previously trained) on the region to recognize shading characteristic of the eyes and the bridge of the nose.

By doing so, the set of possible eye-locations in the image is reduced from the previous level based on a feature filter. This allows the eye detector to run in real-time on a 400 MHz PC. The methodology assumes that the lighting conditions allow the eyes to be distinguished as dark regions surrounded by highlights of the temples and the bridge of the nose, that human eyes are largely surrounded by regions of skin color, that the head is only moderately rotated, that the eyes are reasonably horizontal, and that people are within interaction distance from the robot (3 to 7 feet).

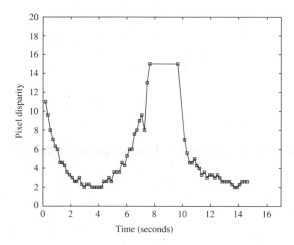

Figure 6.6
This plot illustrates how the target proximity measure varies with distance. The subject begins by standing approximately 2 feet away from the robot ($t = 0$). He then steps back to a distance of about 7 feet ($t = 4$). This is on the outer periphery of the robot's interaction range. Beyond this distance, the robot does not reliably attend to the person as the target of interest as other things are often more salient. The subject then approaches the robot to a distance of 3 inches from its face ($t = 8$ to $t = 10$). The loom detector is firing, which is the plateau in the graph. At $t = 10$ the subject then backs away and leaves the scene.

Proximity estimation Given a target in the visual field, proximity is computed from a stereo match between the two wide cameras. The target in the central wide camera is located within the lower wide camera by searching along epipolar lines for a sufficiently similar patch of pixels, where similarity is measured using normalized cross-correlation. This matching process is repeated for a collection of points around the target to confirm that the correspondences have the right topology. This allows many spurious matches to be rejected. Figure 6.6 illustrates how this metric changes with distance from the robot. It is reasonably monotonic, but subject to noise. It is also quite sensitive to the orientations of the two wide center cameras.

Loom detection The loom calculation makes use of the two cameras with wide fields of view. These cameras are parallel to each other, so when there is nothing in view that is close to the cameras (relative to the distance between them), their output tends to be very similar. A close object, on the other hand, projects very differently on to the two cameras, leading to a large difference between the two views.

   By simply summing the pixel-by-pixel differences between the images from the two cameras, a measure is extracted which becomes large in the presence of a close object. Since Kismet's wide cameras are quite far from each other, much of the room and furniture is close enough to introduce a component into the measure which will change as Kismet looks around. To compensate for this, the measure is subject to rapid habituation. This has the side-effect that a slowly approaching object will not be detected—which is perfectly acceptable for a loom response where the robot quickly withdraws from a sudden and rapidly approaching object.

Threat detection A nearby object (as computed above) along with large but concentrated movement in the wide FoV is treated as a threat by Kismet. The amount of motion corresponds to the amount of activation of the motion map. Since the motion map may also become very active during ego-motion, this response is disabled for the brief intervals during which Kismet's head is in motion. As an additional filtering stage, the ratio of activation in the peripheral part of the image versus the central part is computed to help reduce the number of spurious threat responses due to ego-motion. This filter thus looks for concentrated activation in a localized region of the motion map, whereas self-induced motion causes activation to smear evenly over the map.

6.3   Results and Evaluation

The overall attention system runs at 20 Hz on several 400 MHz processors. In this section, I evaluate its behavior with respect to directing Kismet's attention to task-relevant stimuli. I also examine how easy it is people to direct the robot's attention to a specific target stimulus, and to determine when they have been successful in doing so.

Effect of Gain Adjustment on Saliency

In section 6.1, I described how the active behavior can manipulate the relative contributions of the bottom-up processes to benefit goal achievement. Figure 6.7 illustrates how the skin tone, motion, and color gains are adjusted as a function of drive intensity, the active behavior, and the nature and quality of the perceptual stimulus.

As shown in figure 6.7, when the `social-drive` is activated by face stimuli (middle), the skin-tone gain is influenced by the `seek-people` and `avoid-people` behaviors. The effects on the gains are shown on the left side of the top plot. When the `stimulation-drive` is activated by color stimuli (bottom), the color gain is influenced by the `seek-toys` and `avoid-toys` behaviors. This is shown to the right of the top plot. Seeking people results in enhancing the face gain and avoiding people results in suppressing the face gain. The color gain is adjusted in a similar fashion when toy-oriented behaviors are active (enhancement when seeking out, suppression during avoidance). The middle plot shows how the `social-drive` and the quality of social stimuli determine which people-oriented behavior is activated. The bottom plot shows how the `stimulation-drive` and the quality of toy stimuli determine which toy-oriented behavior is active. All parameters shown in these plots were recorded during the same four-minute period.

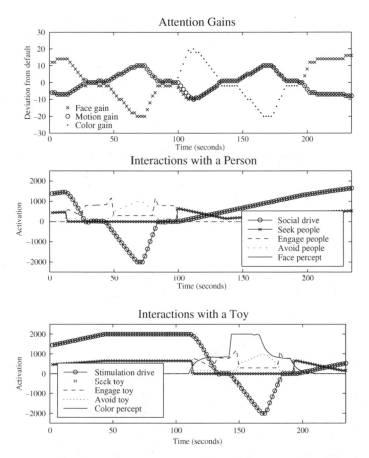

Figure 6.7
Changes of the skin tone, motion, and color gains from top-down motivational and behavioral influences (top). On the left half of the top figure, the gains change with respect to person-related behaviors (middle figure). On the right half of the top figure, the gains change with respect to toy-related behaviors (bottom figure).

The relative weighting of the attention gains are empirically set to satisfy behavioral performance as well as to satisfy social interaction dynamics. For instance, when engaging in visual search, the attention gains are set so that there is a strong preference for the target stimulus (skin tone when searching for social stimuli like people, saturated color when searching for non-social stimuli like toys). As shown in figure 6.3, a distant face has greater overall saliency than a nearby toy if the robot is actively looking for skin-toned stimuli. Similarly, as shown to the right in figure 6.3, a distant toy has greater overall saliency than a nearby face when the robot is actively seeking out stimuli of highly saturated color.

Behaviorally, the robot will continue to search upon encountering a static object of high raw saliency but of the wrong feature. Upon encountering a static object possessing the right saliency feature, the robot successfully terminates search and begins to visually engage the object. However, the search behavior sets the attention gains to allow Kismet to attend to a stimulus possessing the wrong saliency feature if it is also supplemented with motion. Hence, if a person really wants to attract the robot's attention to a specific target that the robot is not actively seeking out, then he/she is still able to do so.

During engagement, the gains are set so that Kismet slightly prefers those stimuli possessing the favored feature. If a stimulus of the favored feature is not present, a stimulus possessing the unfavored feature is sufficient to attract the robot's attention. Thus, while engaged, the robot can satiate other motivations in an opportunistic manner when the desired stimulus is not present. If, however, the robot is unable to satiate a specific motivation for a prolonged time, the motive to engage that stimuli will increase until the robot eventually breaks engagement to preferentially search for the desired stimulus.

Effect of Gain Adjustment on Looking Preference

Figure 6.8 illustrates how top-down gain adjustments combine with bottom-up habituation effects to bias the robot's gaze. When the `seek-people` behavior is active, the skin-tone gain is enhanced and the robot prefers to look at a face over a colorful toy. The robot eventually habituates to the face stimulus and switches gaze briefly to the toy stimulus. Once the robot has moved its gaze away from the face stimulus, the habituation is reset and the robot rapidly reacquires the face. In one set of behavioral trials when `seek-people` was active, the robot spent 80 percent of the time looking at the face. A similar affect can be seen when the `seek-toy` behavior is active—the robot prefers to look at a toy (rather than a face) 83 percent of the time.

The opposite effect is apparent when the `avoid-people` behavior is active. In this case, the skin-tone gain is suppressed so that faces become less salient and are more rapidly affected by habituation. Because the toy is relatively more salient than the face, it takes longer for the robot to habituate. Overall, the robot looks at faces only 5 percent of the time when in this behavioral context. A similar scenario holds when the robot's `avoid-toy` behavior is active—the robot looks at toys only 24 percent of the time.

Socially Manipulating Attention

Figure 6.9 shows an example of the attention system in use, choosing stimuli that are potentially behaviorally relevant in a complex scene. The attention system runs all the time, even when it is not controlling gaze direction, since it determines the perceptual input to which the motivational and behavioral systems respond. Because the robot attends to a

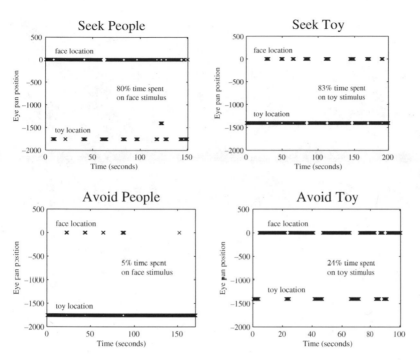

Figure 6.8
Preferential looking based on habituation and top-down influences. These plots illustrate how Kismet's preference for looking at different types of stimuli (a person's face versus a brightly colored toy) varies with top-down behavior and motivational factors.

subset of the same cues that humans find interesting, people naturally and intuitively direct the robot's gaze to a desired target.

Three naive subjects were invited to interact with Kismet. The subjects ranged in age from 25 to 28 years old. All used computers frequently but were not computer scientists by training. All interactions were video-recorded. The robot's attention gains were set to their default values so that there would be no strong preference for one saliency feature over another.

The subjects were asked to direct the robot's attention to each of the target stimuli. There were seven target stimuli used in the study. Three were saturated color stimuli, three were skin-toned stimuli, and the last was a pure motion stimulus. The CD-ROM shows one of the subjects performing this experiment. Each target stimulus was used more than once per subject. These are listed below:

- A highly saturated colorful block
- A bright yellow stuffed dinosaur with multi-color spines

Figure 6.9
Manipulating the robot's attention. Images on the top row are from Kismet's upper wide camera. Images on the bottom summarize the contemporaneous state of the robot's attention system. Brightness in the lower image corresponds to salience; rectangles correspond to regions of interest. The thickest rectangles correspond to the robot's locus of attention. The robot's motivation here is such that stimuli associated with faces and stimuli associated with toys are equally weighted. In the first pair of images, the robot is attending to a face and engaging in mutual regard. By shaking the colored block, its salience increases enough to cause a switch in the robot's attention. The third pair shows that the head and eyes track the toy as it moves, giving feedback to the human as to the robot's locus of attention. In the fourth pair, the robot's attention switches back to the human's face, which is tracked as it moves.

- A bright green cylinder
- A bright pink cup (which is actually detected by the skin tone feature map)
- The person's face
- The person's hand
- A black and white plush cow (which is only salient when moving)

The video was later analyzed to determine which cues the subjects used to attract the robot's attention, which cues they used to determine when they had been successful, and the length of time required to do so. They were also interviewed at the end of the session about which cues they used, which cues they read, and about how long they thought it took to direct the robot's attention. The results are summarized in table 6.1.

To attract the robot's attention, the most frequently used cues include bringing the target close and in front of the robot's face, shaking the object of interest, or moving it slowly across the centerline of the robot's face. Each cue increases the saliency of a stimulus by making it appear larger in the visual field, or by supplementing the color or skin-tone cue with motion. Note that there was an inherent competition between the saliency of the target and the subject's own face as both could be visible from the wide FoV camera. If the subject did not try to direct the robot's attention to the target, the robot tended to look at the subject's face.

Table 6.1
Summary from attention manipulation studies.

Stimulus Category	Stimulus	Presentations	Average Time(s)	Commonly Used Cues	Commonly Read Cues
Color and Movement	Yellow Dinosaur	8	8.5	Motion across center line	Eye behavior, esp. tracking
	Multi-Colored Block	8	6.5		
	Green Cylinder	8	6.0	Shaking motion	Facial expression, esp. raised brows
Motion Only	Black and White Cow	8	5.0		
Skin Tone and Movement	Pink Cup	8	6.5	Bringing target close to robot	Body posture, esp. leaning toward or away
	Hand	8	5.0		
	Face	8	3.5		
Total		56	5.9		

The subjects also effortlessly determined when they had successfully re-directed the robot's gaze. Interestingly, it is not sufficient for the robot to orient to the target. People look for a *change* in visual behavior, from ballistic orientation movements to smooth pursuit movements, before concluding that they had successfully re-directed the robot's attention. All subjects reported that eye movement was the most relevant cue to determine if they had successfully directed the robot's attention. They all reported that it was easy to direct the robot's attention to the desired target. They estimated the mean time to direct the robot's attention at 5 to 10 seconds. This turns out to be the case; the mean time over all trials and all targets is 5.8 seconds.

6.4   Limitations and Extensions

There are a number of ways the current implementation can be improved and expanded upon. Some of these recommendations involve supplementing the existing framework; others involve integrating this system into a larger framework.

One interesting way this system can be improved is by adding a stereo depth map. Currently, the system estimates the proximity of the selected target. A depth map would be very useful as a bottom-up contribution. For instance, regions corresponding to closer proximity to the robot should be more salient than those further away. A stereo map would also be very useful for scene segmentation to separate stimuli of interest from background. This can be accomplished by using the two central wide FoV cameras.

Another interesting feature map to incorporate would be edge orientation. Wolfe, Triesman, and others argue in favor of edge orientation as a bottom-up feature map in humans. Currently, Kismet has no shape metrics to help it distinguish objects from each other (such as its block from its dinosaur). Adding features to support this is an important extension to the existing implementation.

There are no auditory bottom-up contributions. A sound localization feature map would be a nice multi-modal extension (Irie, 1995). Currently, Kismet assumes that the most salient person is the one who is talking to it. Often there are multiple people talking around and to the robot. It is important that the robot knows who is addressing it and when. Sound localization would be of great benefit here. Fortunately, there are stereo microphones on Kismet's ears that could be used for this purpose.

Another interesting extension would be to separate the color saliency map into individual color feature maps. Kismet can preferentially direct its attention to saturated color, but not specifically to green, blue, red, or yellow. Humans are capable of directing search based on a specific color channel. Although Kismet has access to the average r, g, b, y components of the target stimulus, it would be nice if it could keep these colors segmented (so that it can distinguish a blue circle on a green background, for instance). Computing individual color feature maps would be a step towards these extensions.

Currently there is nothing that modifies the decay rate of the habituation feature map. The habituation contribution implements a primitive attention span for the robot. It would be an interesting extension to have motivational factors, such as fatigue or arousal, influence the habituation decay rate. Caregivers continually adjust the arousal level of their infant so that the infant remains alert but not too excited (Bullowa, 1979). For Kismet, it would be interesting if the human could adjust the robot's attention span by keeping it at a moderate arousal level. This could benefit the robot's learning rate by maintaining a longer attention span when people are around and the robot is engaged in interactions with high learning potential.

Kismet's visual perceptual world consists only of what is in view of the cameras. Ultimately, the robot should be able to construct an ego-centered saliency map of interaction space. In this representation, the robot could keep track of where interesting things are located, even if they are not currently in view. This will prove to be a very important representation for social referencing (Siegel, 1999). If Kismet could engage in social referencing, then it could look to the human for the affective assessment and then back to the event that it queried the caregiver about. Chances are, the event in question and the human's face will not be in view at the same time. Hence, a representation of where interesting things are, even when out of view, is an important resource.

6.5 Summary

There are many interesting ways in which Kismet's attention system can be improved and extended. This should not overshadow the fact that the existing attention system is an important contribution to autonomous robotics research.

Other researchers have developed bottom-up attention systems (Itti et al., 1998; Wolfe, 1994). Many of these systems work in isolation and are not embedded in a behaving robot. Kismet's attention system goes beyond raw perceptual saliency to incorporate top-down task-driven influences that vary dynamically over time with its goals. By doing so, the attention system is tuned to benefit the task the robot is currently engaged in.

There are far too many things that the robot could be responding to at any time. The attention system gives the robot a locus of interest that it can organize its behavior around. This contributes to perceptual stability, since the robot is not inclined to flit its eyes around randomly from place to place, changing its perceptual input at a pace too rapid for behavior to keep up. This in turn contributes to behavioral stability since the robot has a target that it can direct its behavior toward and respond to. Each target (people, toys) has a physical persistence that is well-matched to the robot's behavioral time scale. Of course, the robot can respond to different targets sequentially in time, but this occurs at a slow enough time scale that the behaviors have time to self-organize and stabilize into a coherent goal-directed pattern before a switch to a new behavior is made.

There is no prior art in incorporating a task-dependent attentional system into a robot. Some sidestep the issue by incorporating an implicit attention mechanism into the perceptual conditions that release behaviors (Blumberg, 1994; Velasquez, 1998). Others do so by building systems that are hardwired to perceive one type of stimulus tailored to the specific task (Schaal, 1997; Mataric et al., 1998), or use very simple sensors (Hayes & Demiris, 1994; Billard & Dautenhahn, 1997). However, the complexity of Kismet's visual environment, the richness of its perceptual capabilities, and its time-varying goals required an explicit implementation.

The social dimension of Kismet's world adds additional constraints that prior robotic systems have not had to deal with. As argued earlier, the robot's attention system must be tuned to the attention system of humans. In this way, both robot and humans are more likely to find the same sorts of things interesting or attention-grabbing. As a result, people can very naturally and quickly direct the robot's attention. The attention system coupled with gaze direction provides people with a powerful and intuitive social cue. The readability and interpretation of the robot's behavior is greatly enhanced since the person has an accurate measure of what the robot is responding to.

The ability for humans to easily influence the robot's attention and to read its cues has a tremendous benefit to various forms of social learning and is an important form of scaffolding. When learning a task, it is difficult for a robotic system to learn what perceptual aspects matter. This only gets worse as robots are expected to perform more complex tasks in more complex environments. This challenging learning issue can be addressed in an interesting way, however, if the robot learns the task with a human instructor who can explicitly direct the robot's attention to the salient aspects and who can determine from the robot's social cues whether or not the robot is attending to the relevant features. This doesn't solve the problem, but it could facilitate a solution in a new and interesting way that is natural and intuitive for people.

In the big picture, low-level feature extraction and visual attention are components of a larger visual system. I present how the attention system is integrated with other visual behaviors in chapter 12.

7 The Auditory System

Human speech provides a natural and intuitive interface both for communicating with and teaching humanoid robots. In general, the acoustic pattern of speech contains three kinds of information: who the speaker is, what the speaker said, and how the speaker said it. This chapter focuses on the problem of recognizing affective intent in robot-directed speech. The work presented in this chapter was carried out in collaboration with Lijin Aryananda (Breazeal & Aryananda, 2002).

When extracting the affective message of a speech signal, there are two related yet distinct questions one can ask. The first: *"What emotion is being expressed?"* In this case, the answer describes an emotional quality—such as sounding angry, or frightened, or disgusted. Each emotional state causes changes in the autonomic nervous system. This, in turn, influences heart rate, blood pressure, respiratory rate, sub-glottal pressure, salivation, and so forth. These physiological changes produce global adjustments to the acoustic correlates of speech—influencing pitch, energy, timing, and articulation. There have been a number of vocal emotion recognition systems developed in the past few years that use different variations and combinations of those acoustic features with different types of learning algorithms (Dellaert et al., 1996; Nakatsu et al., 1999). To give a rough sense of performance, a five-way classifier operating at approximately 80 percent is considered state of the art (at the time of this writing). This is impressive considering that humans are far from perfect in recognizing emotion from speech alone. Some have attempted to use multi-modal cues (facial expression with expressive speech) to improve recognition performance (Chen & Huang, 1998).

7.1 Recognizing Affect in Human Speech

For the purposes of training a robot, however, the raw emotional content of the speaker's voice is only part of the message. This leads us to the second, related question: *What is the affective intent of the message?* Answers to this question may be that the speaker was praising, prohibiting, or alerting the recipient of the message. A few researchers have developed systems that can recognize speaker approval versus speaker disapproval from child-directed speech (Roy & Pentland, 1996), or recognize praise, prohibition, and attentional bids from infant-directed speech (Slaney & McRoberts, 1998). For the remainder of this chapter, I discuss how this idea could be extended to serve as a useful training signal for Kismet. Note that Kismet does not learn from humans yet, but this is an important capability that could support socially situated learning.

Developmental psycholinguists have extensively studied how affective intent is communicated to preverbal infants (Fernald, 1989; Grieser & Kuhl, 1988). Infant-directed speech is typically quite exaggerated in pitch and intensity (Snow, 1972). From the results of a series of cross-cultural studies, Fernald suggests that much of this information is communicated through the "melody" of infant-directed speech. In particular, there is evidence for at least four distinctive prosodic contours, each of which communicates a different affective meaning to the infant (approval, prohibition, comfort, and attention). Maternal exaggerations in infant-directed speech seem to be particularly well-matched to the innate affective responses of human infants (Mumme et al., 1996).

Inspired by this work, Kismet uses a recognizer to distinguish the four affective intents for praise, prohibition, comfort, and attentional bids. Of course, not everything a human says to Kismet will have an affective meaning, so neutral robot-directed speech is also distinguished. These affective intents are well-matched to teaching a robot since praise (positive reinforcement), prohibition (negative reinforcement), and directing attention could be intuitively used by a human instructor to facilitate the robot's learning process. Within the AI community, a few researchers have already demonstrated how affective information can be used to bias learning at both goal-directed and affective levels for robots (Velasquez, 1998) and synthetic characters (Yoon et al., 2000).

For Kismet, the output of the vocal classifier is interfaced with the emotion subsystem (see chapter 8), where the information is appraised at an affective level and then used to directly modulate the robot's own affective state.[1] In this way, the affective meaning of the utterance is communicated to the robot through a mechanism similar to the one Fernald suggests. As with human infants, socially manipulating the robot's affective system is a powerful way to modulate the robot's behavior and to elicit an appropriate response.

In the rest of this chapter, I discuss previous work in recognizing emotion and affective intent in human speech. I discuss Fernald's work in depth to highlight the important insights it provides in terms of which cues are the most useful for recognizing affective intent, as well as how it may be used by human infants to organize their behavior. I then outline a series of design issues for integrating this competence into Kismet. I present a detailed description of the approach implemented on Kismet and how it has been integrated into Kismet's affective circuitry. The performance of the system is evaluated with naive subjects as well as the robot's caregivers. I discuss the results, suggest future work, and summarize findings.

7.2   Affect and Meaning in Infant-Directed Speech

Developmental psycholinguists have studied the acoustic form of adult speech directed to preverbal infants and have discovered an intriguing relation between voice pitch and affective intent (Fernald, 1989; Papousek et al., 1985; Grieser & Kuhl, 1988). When mothers

---

1. Typically, "affect" refers to positive and negative qualities. For Kismet, arousal levels and the robot's willingness to approach or withdraw are also included when talking about Kismet's affective state.

speak to their preverbal infant, their prosodic patterns (the contour of the fundamental frequency and modulations in intensity) are exaggerated in characteristic ways. Even with newborns, mothers use higher mean pitch, wider pitch range, longer pauses, shorter phrases, and more prosodic repetition when addressing infants than when speaking to an adult. These affective contours have been found to exist in several cultures. This exaggerated manner of speaking (i.e., motherese) serves to engage infant's attention and prolong interaction.

Maternal intonation is finely tuned to the behavioral and affective state of the infant. Further, mothers intuitively use selective prosodic contours to express different affective intentions, most notably those for praise, prohibition, soothing, and attentional bids. Based on a series of cross-linguistic analyses, there appear to be at least four different pitch contours (approval, prohibition, comfort, and attentional bids), each associated with a different emotional state (Grieser & Kuhl, 1988; Fernald, 1993; McRoberts et al., 2000). Mothers are more likely to use falling pitch contours than rising pitch contours when soothing a distressed infant (Papousek et al., 1985), to use rising contours to elicit attention and to encourage a response (Ferrier, 1985), and to use bell-shaped contours to maintain attention once it has been established (Stern et al., 1982). Expressions of approval or praise, such as "Good girl!" are often spoken with an exaggerated rise-fall pitch contour with sustained intensity at the contour's peak. Expressions of prohibitions or warnings such as "Don't do that!" are spoken with low pitch and high intensity in staccato pitch contours. Figure 7.1 illustrates these prototypical contours.

It is interesting that even though preverbal infants do not understand the linguistic content of the message, they appear to understand the affective content and respond appropriately. It seems that the exaggerated prosodic cues convey meaning. This may comprise some of infants' earliest communicated meanings of maternal vocalizations. The same patterns can be found when communicating these same intents to adults, but in a significantly less exaggerated manner (Fernald, 1989). By eliminating the linguistic content of

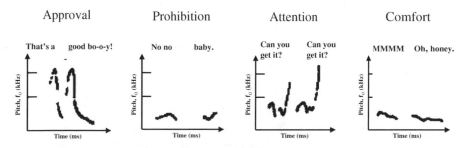

Figure 7.1
Fernald's prototypical contours for approval, prohibition, attention, and soothing. It is argued that they are well-matched to saliency measures hardwired into an infant's auditory processing system.

infant-directed and adult-directed utterances for the categories described above (only preserving the "melody" of the message), Fernald found that adult listeners were more accurate
in recognizing these affective categories in infant-directed speech than in adult-directed
speech. This suggests that the relation of prosodic form to communicative function is made
uniquely salient in the melodies of mother's speech, and that these intonation contours
provide the listener with reliable acoustic cues to the speaker's intent.

Fernald has used the results of such studies to argue for the adaptive significance of
prosody in child language acquisition, as well as in the development and strength of the
parent-offspring relationship. Caregivers are very good at matching the acoustic structure of
their speech to communicative function. Fernald suggests that the pitch contours observed
have been designed to directly influence the infant's emotive state, causing the child to
relax or become more vigilant in certain situations, and to either avoid or approach objects
that may be unfamiliar. Auditory signals with high frequency and rising pitch are more
likely to alert human listeners than signals lower in frequency and falling pitch (Ferrier,
1985). Hence, the acoustic design of attentional bids would appear to be appropriate to the
goal of eliciting attention. Similarly, low mean pitch, narrow pitch range, and low intensity
(all characteristics of comfort vocalizations) have been found to be correlated with low
arousal (Papousek et al., 1985). Given that the mother's goal in soothing her infant is to
decrease arousal, comfort vocalizations are well-suited to this function. Speech having a
sharp, loud, staccato contour, low pitch mean, and narrow pitch range tend to startle the
infant (tending to halt action or even induce withdraw) and are particularly effective as
warning signals (Fernald, 1989). Infants show a listening preference for exaggerated pitch
contours. They respond with more positive affect to wide range pitch contours than to
narrow range pitch contours. The exaggerated bell-shaped prosody contour for approval is
effective for sustaining the infant's attention and engagement (Stern et al., 1982).

By anchoring the message in the melody, there may be a facilitative effect on "pulling"
the word out of the acoustic stream and causing it to be associated with an object or
event. This development is argued to occur in four stages. To paraphrase Fernald (1989),
in the first stage, certain acoustic features of speech have intrinsic perceptual salience for
the infant. Certain maternal vocalizations function as unconditioned stimuli in alerting,
soothing, pleasing, and alarming the infant. In stage two, the melodies of maternal speech
become increasingly more effective in directing the infant's attention and in modulating the
infant's arousal and affect. The communication of intention and emotion takes place in the
third stage. Vocal and facial expressions give the infant initial access to the feelings and
intentions of others. Stereotyped prosodic contours occurring in specific affective contexts
come to function as the first regular sound-meaning correspondences for the infant. In the
fourth stage, prosodic marking of focused words helps the infant to identify linguistic units
within the stream of speech. Words begin to emerge from the melody.

7.3   Design Issues for Recognizing Affective Intent

There are several design issues that must be addressed to successfully integrate Fernald's ideas into a robot like Kismet. As I have argued previously, this could provide a human caregiver with a natural and intuitive means for communicating with and training a robotic creature. The initial communication is at an affective level, where the caregiver socially manipulates the robot's affective state. For Kismet, the affective channel provides a powerful means for modulating the robot's behavior.

Robot aesthetics   As discussed above, the perceptual task of recognizing affective intent is significantly easier in infant-directed speech than in adult-directed speech. Even human adults have a difficult time recognizing intent from adult-directed speech without the linguistic information. It will be a while before robots have true natural language, but the affective content of the vocalization can be extracted from prosody. Encouraging speech on an infant-directed level places a constraint on how the robot appears physically (chapter 5), how it moves (chapters 9, 12), and how it expresses itself (chapters 10, 11). If the robot looks and behaves as a very young creature, people will be more likely to treat it as such and naturally exaggerate their prosody when addressing the robot. This manner of robot-directed speech would be spontaneous and seem quite appropriate. I have found this typically to be the case for both men and women when interacting with Kismet.

Real-time performance   Another design constraint is that the robot be able to interpret the vocalization and respond to it at natural interactive rates. The human can tolerate small delays (perhaps a second or so), but long delays will break the natural flow of the interaction. Long delays also interfere with the caregiver's ability to use the vocalization as a reinforcement signal. Given that the reinforcement should be used to mark a specific event as good or bad, long delays could cause the wrong action to be reinforced and confuse the training process.

Voice as training signal   People should be able to use their voice as a natural and intuitive training signal for the robot. The human voice is quite flexible and can be used to convey many different meanings, affective or otherwise. The robot should be able to recognize when it is being praised and associate it with positive reinforcement. Similarly, the robot should recognize scolding and associate it with negative reinforcement. The caregiver should be able to acquire and direct the robot's attention with attentional bids to the relevant aspects of the task. Comforting speech should be soothing for the robot if it is in a distressed state, and encouraging otherwise.

Voice as saliency marker   This raises a related issue, which is the caregiver's ability to use their affective speech as a means of marking a particular event as salient. This implies that the robot should *only* recognize a vocalization as having affective content in the cases where the caregiver specifically intends to praise, prohibit, soothe, or get the attention of the robot. The robot should be able to recognize neutral robot-directed speech, even if it is somewhat tender or friendly in nature (as is often the case with motherese). For this reason, the recognizer only categorizes sufficiently exaggerated prosody such as as praise, prohibition, attention, and soothing (i.e., the caregiver has to say it as if she *really* means it). Vocalizations with insufficient exaggeration are classified as neutral.

Acceptable versus unacceptable misclassification   Given that humans are not perfect at recognizing the affective content in speech, the robot is sure to make mistakes as well. However, some failure modes are more acceptable than others. For a teaching task, confusing strongly valenced intent for neutrally valenced intent is better than confusing oppositely valenced intents. For instance, confusing approval for an attentional bid, or prohibition for neutral speech, is better than interpreting prohibition for praise. Ideally, the recognizer's failure modes will minimize these sorts of errors.

Expressive feedback   Nonetheless, mistakes in communication will be made. This motivates the need for feedback from the robot back to the caregiver. Fundamentally, the caregiver is trying to communicate his/her intent to the robot. The caregiver has no idea whether or not the robot interpreted the intent correctly without some form of feedback. By interfacing the output of the recognizer to Kismet's emotional system, the robot's ability to express itself through facial expression, voice quality, and body posture conveys the robot's affective interpretation of the message. This allows people to reiterate themselves until they believe they have been properly understood. It also enables the caregiver to reiterate the message until the intent is communicated strongly enough (perhaps what the robot just did was *very* good, and the robot should be *really* happy about it).

Speaker dependence versus independence   An interesting question is whether the recognizer should be speaker-dependent or speaker-independent. There are obviously advantages and disadvantages to both, and the appropriate choice depends on the application. Typically, it is easier to get higher recognition performance from a speaker-dependent system. In the case of a personal robot, this is a good alternative since the robot should be personalized to a particular human over time, not preferentially tuned to others. If the robot must interact with a wide variety of people, then the speaker-independent system is preferable. The underlying question in both cases is what level of performance is necessary for people to feel that the robot is responsive and understands them well enough so that it is not challenging or frustrating to communicate with it and train it.

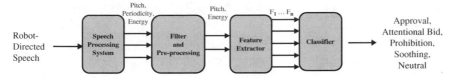

Figure 7.2
The spoken affective intent recognizer.

7.4   The Affective Intent Classifier

As shown in figure 7.2, the affective speech recognizer receives robot-directed speech as input. The speech signal is analyzed by the low-level speech processing system, producing time-stamped pitch (Hz), percent periodicity (a measure of how likely a frame is a voiced segment), energy (dB), and phoneme values[2] in real-time. The next module performs filtering and pre-processing to reduce the amount of noise in the data. The pitch value of a frame is simply set to 0 if the corresponding percent periodicity indicates that the frame is more likely to correspond to unvoiced speech. The resulting pitch and energy data are then passed through the feature extractor, which calculates a set of selected features (F_1 to F_n). Finally, based on the trained model, the classifier determines whether the computed features are derived from an approval, an attentional bid, a prohibition, soothing speech, or a neutral utterance.

Two female adults who frequently interact with Kismet as caregivers were recorded. The speakers were asked to express all five affective intents (approval, attentional bid, prohibition, comfort, and neutral) during the interaction. Recordings were made using a wireless microphone, and the output signal was sent to the low-level speech processing system running on Linux. For each utterance, this phase produced a 16-bit single channel, 8 kHz signal (in a .wav format) as well as its corresponding real-time pitch, percent periodicity, energy, and phoneme values. All recordings were performed in Kismet's usual environment to minimize variability of environment-specific noise. Samples containing extremely loud noises (door slams, etc.) were eliminated, and the remaining data set were labeled according to the speakers' affective intents during the interaction. There were a total of 726 utterances in the final data set—approximately 145 utterances per class.

The pitch value of a frame was set to 0 if the corresponding percent periodicity was lower than a threshold value. This indicates that the frame is more likely to correspond

2. This auditory processing code is provided by the Spoken Language Systems Group at MIT. For now, the phoneme information is not used in the recognizer.

to unvoiced speech. Even after this procedure, observation of the resulting pitch contours still indicated the presence of substantial noise. Specifically, a significant number of errors were discovered in the high pitch value region (above 500 Hz). Therefore, additional preprocessing was performed on all pitch data. For each pitch contour, a histogram of ten regions was constructed. Using the heuristic that the pitch contour was relatively smooth, it was determined that if only a few pitch values were located in the high region while the rest were much lower (and none resided in between), then the high values were likely to be noise. Note that this process did not eliminate high but smooth pitch contour since pitch values would be distributed evenly across nearby regions.

Classification Method

In all training phases each class of data was modeled using a Gaussian mixture model, updated with the EM algorithm and a Kurtosis-based approach for dynamically deciding the appropriate number of kernels (Vlassis & Likas, 1999). Due to the limited set of training data, cross-validation in all classification processes was performed. Specifically, a subset of data was set aside to train a classifier using the remaining data. The classifier's performance was then tested on the held-out test set. This process was repeated 100 times per classifier. The mean and variance of the percentage of correctly classified test data were calculated to estimate the classifier's performance.

As shown in figure 7.3, the preprocessed pitch contour in the labeled data resembles Fernald's prototypical prosodic contours for approval, attention, prohibition, and comfort/ soothing. A set of global pitch and energy related features (see table 7.1) were used to recognize these proposed patterns. All pitch features were measured using only non-zero pitch values. Using this feature set, a sequential forward feature selection process was applied to construct an optimal classifier. Each possible feature pair's classification performance was measured and sorted from highest to lowest. Successively, a feature pair from the sorted list was added into the selected feature set to determine the best n features for an optimal classifier. Table 7.2 shows the results of the classifiers constructed using the best eight feature pairs. Classification performance increases as more features are added, reaches maximum (78.77 percent) with five features in the set, and levels off above 60 percent with six or more features. It was found that global pitch and energy measures were useful in roughly separating the proposed patterns based on arousal (largely distinguished by energy measures) and valence (largely distinguished by pitch measures). However, further processing was required to distinguish each of the five classes distinctly.

Accordingly, the classifier consists of several mini-classifiers executing in stages. In the beginning stages, the classifier uses global pitch and energy features to separate some of the classes into pairs (in this case, clusters of soothing along with low-energy neutral, prohibition along with high-energy neutral, and attention along with approval were formed).

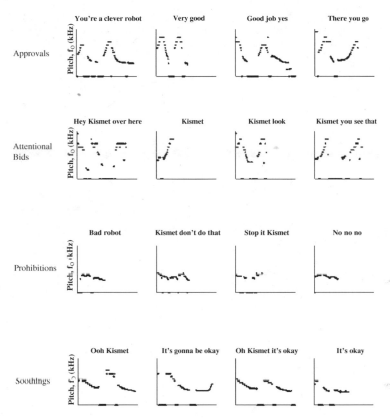

Figure 7.3
Fernald's prototypical prosodic contours found in the preprocessed data set. Notice the similarity to those shown in figure 7.1.

These clustered classes were then passed to additional classification stages for further refinement. New features had to be considered to build these additional classifiers. Using prior information, a new set of features encoding the shape of the pitch contour was included, which proved useful in further separating the classes.

To select the best features for the initial classification stage, the seven feature pairs listed in table 7.2 were examined. All feature pairs worked better in separating prohibition and soothing than other classes. The F_1-F_9 pair generates the highest overall performance and the least number of errors in classifying prohibition. Several observations can be made from the feature space of this classifier (see figure 7.4). The prohibition samples are clustered in the low pitch mean and high energy variance region. The approval and attention classes form a cluster at the high pitch mean and high energy variance region. The soothing

Table 7.1
Features extracted in the first-stage classifier. These features are measured over the non-zero values throughout the entire utterance. Feature F_6 measures the steepness of the slope of the pitch contour.

Feature	Description
F_1	Pitch mean
F_2	Pitch Variance
F_3	Maximum Pitch
F_4	Minimum Pitch
F_5	Pitch Range
F_6	Delta Pitch Mean
F_7	Absolute Delta Pitch Mean
F_8	Energy Mean
F_9	Energy Variance
F_{10}	Energy Range
F_{11}	Maximum Energy
F_{12}	Minimum Energy

Table 7.2
The performance (the percent correctly classified) is shown for the best pair-wise set having up to eight features. The pair-wise performance was ranked for the best seven pairs. As each successive feature was added, performance peaks with five features (78.8%), but then drops off.

Feature Pair	Feature Set	Perf. Mean	Perf. Variance	Percent Error Approval	Percent Error Attention	Percent Error Prohibition	Percent Error Soothing	Percent Error Neutral
F_1, F_9	$F_1 F_9$	72.1	0.1	48.7	24.5	8.7	15.6	42.1
F_1, F_{10}	$F_1 F_9 F_{10}$	75.2	0.1	41.7	25.7	9.7	13.2	34.0
F_1, F_{11}	$F_1 F_9 F_{10} F_{11}$	78.1	0.1	29.9	27.2	8.8	10.6	34.0
F_2, F_9	$F_1 F_2 F_9 F_{10} F_{11}$	78.8	0.1	29.2	22.2	8.5	12.6	33.7
F_3, F_9	$F_1 F_2 F_3 F_9 F_{10} F_{11}$	61.5	1.2	63.9	43.0	9.1	23.1	53.4
F_1, F_8	$F_1 F_2 F_3 F_8 F_9 F_{10} F_{11}$	62.3	1.8	60.6	39.6	16.4	24.2	47.9
F_5, F_9	$F_1 F_2 F_3 F_5 F_8 F_9 F_{10} F_{11}$	65.9	0.7	57.0	32.2	12.1	19.7	49.4

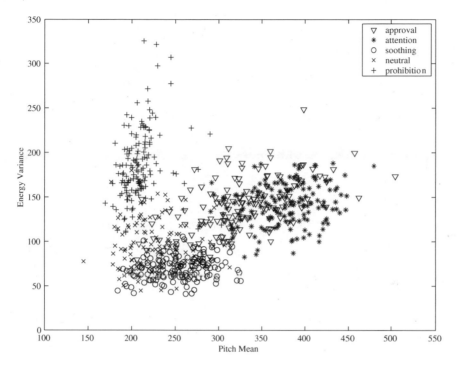

Figure 7.4
Feature space of all five classes with respect to energy variance, F_9, and pitch mean, F_1. There are three distinguishable clusters for prohibition, soothing and neutral, and approval and attention.

samples are clustered in the low pitch mean and low energy variance region. The neutral samples have low pitch mean and are divided into two regions in terms of their energy variance values. The neutral samples with high energy variance are clustered separately from the rest of the classes (in between prohibition and soothing), while the ones with lower energy variance are clustered within the soothing class. These findings are consistent with the proposed prior knowledge. Approval, attention, and prohibition are associated with high intensity while soothing exhibits much lower intensity. Neutral samples span from low to medium intensity, which makes sense because the neutral class includes a wide variety of utterances.

Based on this observation, the first classification stage uses energy-related features to classify soothing and low-intensity neutral with from the other higher intensity classes (see figure 7.5). In the second stage, if the utterance had a low intensity level, another classifier decides whether it is soothing or neutral. If the utterance exhibited high intensity, the F_1-F_9 pair is used to classify among prohibition, the approval-attention cluster, and high intensity

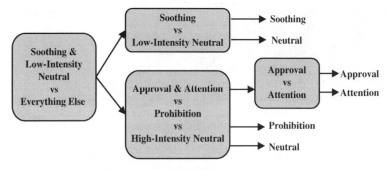

Figure 7.5
The classification stages of the multi-stage classifier.

Table 7.3
Classification results in stage 1.

Feature Pair	Pair Perf. Mean (%)	Feature Set	Perf. Mean (%)
F_9, F_{11}	93.0	$F_9 F_{11}$	93.0
F_{10}, F_{11}	91.8	$F_9 F_{10} F_{11}$	93.6
F_2, F_9	91.7	$F_2 F_9 F_{10} F_{11}$	93.3
F_7, F_9	91.3	$F_2 F_7 F_9 F_{10} F_{11}$	91.6

neutral. An additional stage is required to classify between approval and attention if the utterance happened to fall within the approval-attention cluster.

Stage 1: Soothing—low-intensity neutral versus everything else   The first two columns in table 7.3 show the classification performance of the top four feature pairs (sorted according to how well each pair classifies soothing and low-intensity neutral against other classes). The last two columns illustrate the classification results as each pair is added sequentially into the feature set. The final classifier was constructed using the best feature set (energy variance, maximum energy, and energy range), with an average performance of 93.6 percent.

Stage 2A: Soothing versus low-intensity neutral   Since the global and energy features were not sufficient in separating these two classes, new features were introduced into the classifier. Fernald's prototypical prosodic patterns for soothing suggest looking for a smooth pitch contour exhibiting a frequency down-sweep. Visual observations of the neutral samples in the data set indicated that neutral speech generated flatter and choppier pitch contours as well as less-modulated energy contours. Based on these postulations, a classifier using five features (number of pitch segments, average length of pitch segments, minimum length of pitch segments, slope of pitch contour, and energy range) was constructed. The slope of the pitch contour indicated whether the contour contained a down-sweep segment. It was calculated by performing a linear fit on the contour segment starting at the maximum peak. This classifier's average performance is 80.3 percent.

Stage 2B: Approval-attention versus prohibition versus high-intensity neutral A combination of pitch mean and energy variance works well in this stage. The resulting classifier's average performance is 90.0 percent. Based on Fernald's prototypical prosodic patterns, it was speculated that pitch variance would be a useful feature for distinguishing between prohibition and the approval-attention cluster. Adding pitch variance into the feature set increased the classifier's average performance to 92.1 percent.

Stage 3: Approval versus attention Since the approval class and attention class span the same region in the global pitch versus energy feature space, prior knowledge (provided by Fernald's prototypical prosodic contours) gave the basis to introduce a new feature. As mentioned above, approvals are characterized by an exaggerated rise-fall pitch contour. This particular pitch pattern proved useful in distinguishing between the two classes. First, a three-degree polynomial fit was performed on each pitch segment. Each segment's slope sequence was analyzed for a positive slope followed by a negative slope with magnitudes higher than a threshold value. The longest pitch segment that contributed to the rise-fall pattern (which was 0 if the pattern was non-existent) was recorded. This feature, together with pitch variance, was used in the final classifier and generated an average performance of 70.5 percent. Approval and attention are the most difficult to classify because both classes exhibit high pitch and intensity. Although the shape of the pitch contour helped to distinguish between the two classes, it is very difficult to achieve high classification performance without looking at the linguistic content of the utterance.

Overall Classification Performance

The final classifier was evaluated using a new test set generated by the same female speakers, containing 371 utterances. Because each mini-classifier was trained using different portions of the original database (for the single-stage classifier), a new data set was gathered to ensure that no mini-classifier stage was tested on data used to train it. Table 7.4 shows the resulting classification performance and compares it to an instance of the cross-validation results of the best single-stage five-way classifier obtained using the five features described in section 7.4. Both classifiers perform very well on prohibition utterances. The multi-stage classifier performs significantly better in classifying the *difficult* classes, i.e., approval versus attention and soothing versus neutral. This verifies that the features encoding the shape of the pitch contours (derived from prior knowledge provided by Fernald's prototypical prosodic patterns) were very useful.

Table 7.4
Overall classification performance.

Category	Test Size	Classified Approvals	Classified Attentional Bids	Classified Prohibitions	Classified Soothings	Classified Neutrals	% Correctly Classified
Approval	84	64	15	0	5	0	76.2
Attention	77	21	55	0	0	1	74.3
Prohibition	80	0	1	78	0	1	97.5
Soothing	68	0	0	0	55	13	80.9
Neutral	62	3	4	0	3	52	83.9
All	371						81.9

It is important to note that both classifiers produce acceptable failure modes (i.e., strongly valenced intents are incorrectly classified as neutrally valenced intents and not as oppositely valenced ones). All classes are sometimes incorrectly classified as neutral. Approval and attentional bids are generally classified as one or the other. Approval utterances are occasionally confused for soothing and vice versa. Only one prohibition utterance was incorrectly classified as an attentional bid, which is acceptable. The single-stage classifier made one unacceptable error of confusing a neutral utterance as a prohibition. In the multi-stage classifier, some neutral utterances are classified as approval, attention, and soothing. This makes sense because the neutral class covers a wide variety of utterances.

7.5   Integration with the Emotion System

The output of the recognizer is integrated into the rest of Kismet's synthetic nervous system as shown in figure 7.6. Please refer to chapter 8 for a detailed description of the design of the emotion system. In this chapter, I briefly present only those aspects of the emotion system as they are related to integrating recognition of vocal affective intent into Kismet. In the following discussion, I distinguish human emotions from the computational models of emotion on Kismet by the following convention: normal font is used when "emotion" is used as a adjective (such as in emotive responses), boldface font is used when referring to a computational process (such as the `fear` process), and quotes are used when making an analogy to animal or human emotions.

The entry point for the classifier's result is at the auditory perceptual system. Here, it is fed into an associated releaser process. In general, there are many different kinds of releasers defined for Kismet, each combining different contributions from a variety of perceptual and motivational systems. Here, I only discuss those releasers related to the input from the vocal classifier. The output of each vocal affect releaser represents its perceptual contribution to

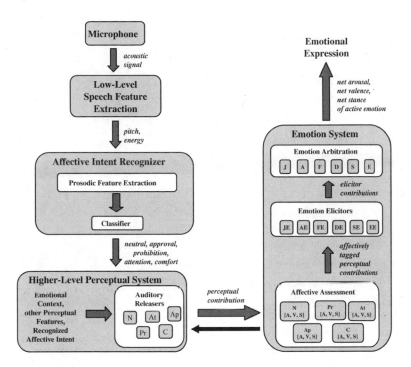

Figure 7.6
System architecture for integrating vocal classifier input to Kismet's emotion system. For the auditory releasers:
N — neutral, Pr — prohibition, At = attention, Ap = approval, and C = comfort. In the emotion system: J
stands for "joy," A stands for "anger," F stands for "fear," D stands for "disgust," S stands for "sorrow," and E
stands for "excited/surprise."

the rest of the SNS. Each releaser combines the incoming recognizer signal with contextual
information (such as the current "emotional" state) and computes its level of activation
according to the magnitude of its inputs. If its activation passes above threshold, it passes
its output on to the emotion system.

Within the emotion system, the output of each releaser must first pass through the
affective assessment subsystem in order to influence "emotional" behavior. Within this assessment subsystem, each releaser is evaluated in affective terms by an associated
somatic marker (SM) process. This mechanism is inspired by the *Somatic Marker Hypothesis* of (Damasio, 1994) where incoming perceptual information is "tagged" with affective
information. Table 7.5 summarizes how each vocal affect releaser is somatically tagged.

There are three classes of tags that the affective assessment phase uses to characterize
its perceptual, motivational, and behavioral input. Each tag has an associated intensity
that scales its contribution to the overall affective state. The *arousal* tag, A, specifies how

Table 7.5
Table mapping [A, V, S] to classified affective intents. Praise biases the robot to be "happy," prohibition biases it to be "sad," comfort evokes a "content, relaxed" state, and attention is "arousing."

Category	Arousal	Valence	Stance	Typical Expression
Approval	medium high	high positive	approach	pleased
Prohibition	low	high negative	withdraw	sad
Comfort	low	medium positive	neutral	content
Attention	high	neutral	aproach	interest
Neutral	neutral	neutral	neutral	calm

arousing this percept is to the emotional system. Positive values correspond to a high arousal stimulus whereas negative values correspond to a low arousal stimulus. The *valence* tag, V, specifies how good or bad this percept is to the emotional system. Positive values correspond to a pleasant stimulus whereas negative values correspond to an unpleasant stimulus. The *stance* tag, S, specifies how approachable the percept is. Positive values correspond to advance whereas negative values correspond to retreat. Because there are potentially many different kinds of factors that modulate the robot's affective state (e.g., behaviors, motivations, perceptions), this tagging process converts the myriad of factors into a common currency that can be combined to determine the net affective state.

For Kismet, the [A, V, S] trio is the currency the emotion system uses to determine which emotional response should be active. This occurs in two phases: First, all somatically marked inputs are passed to the *emotion elicitor* stage. Each `emotion` process has an elicitor associated with it that filters each of the incoming [A, V, S] contributions. Only those contributions that satisfy the [A, V, S] criteria for that `emotion` process are allowed to contribute to its activation. This filtering is done independently for each class of affective tag. Given all these factors, each elicitor computes its net [A, V, S] contribution and activation level, and passes them to the associated `emotion` process within the *emotion arbitration subsystem*. In the second stage, the `emotion` processes within this subsystem compete for activation based on their activation level. There is an `emotion` process for each of Ekman's six basic emotions (Ekman, 1992). The "Ekman six" encompass `joy`, `anger`, `disgust`, `fear`, `sorrow`, and `surprise`. He posits that these six emotions are innate in humans, and all others are acquired through experience.

If the activation level of the winning `emotion` process passes above threshold, it is allowed to influence the behavior system and the motor expression system. There are actually two threshold levels, one for expression and one for behavior. The expression threshold is lower than the behavior threshold; this allows the facial expression to *lead* the behavioral response. This enhances the readability and interpretation of the robot's behavior for the human observer. For instance, given that the caregiver makes an attentional bid, the robot's face will first exhibit an aroused and interested expression, then the orienting response ensues. By staging the response in this manner, the caregiver gets immediate expressive feedback that the robot understood her intent. For Kismet, this feedback can come in a combination of facial expression and posture (chapter 10), or tone of voice (chapter 11). The robot's facial expression also sets up the human's expectation of what behavior will soon follow. As a result, the human observing the robot can see its behavior, in addition to having an understanding of why the robot is behaving in that manner. As I have argued previously, readability is an important issue for social interaction with humans.

Socio-Emotional Context Improves Interpretation

Most affective speech recognizers are not integrated into robots equipped with emotion systems that are also embedded in a social environment. As a result, they have to classify each utterance in isolation. For Kismet, however, the surrounding social context can be exploited to help reduce false categorizations, or at least to reduce the number of "bad" misclassifications (such as mixing up prohibitions for approvals).

Some of this contextual filtering is performed by the transition dynamics of the `emotion` processes. These processes cannot instantaneously become active or inactive. Decay rates and competition for activation with other `emotion` processes give the currently active process a base level of persistence before it becomes inactive. Hence, for a sequence of approvals where the activation of the robot's `joy` process is very high, an isolated prohibition will not be sufficient to immediately switch the robot to a negatively valenced state.

If the caregiver intends to communicate disapproval, reiteration of the prohibition will continue to increase the contribution of negative valence to the emotion system. This serves to inhibit the positively valenced `emotion` processes and to excite the negatively valenced `emotion` processes. Expressive feedback from the robot is sufficient for the caregiver to recognize when the intent of the vocalization has been communicated properly and strongly enough. The smooth transition dynamics of the emotion system enhances the naturalness of the robot's behavior since a person would expect to have to "build up" to a dramatic shift in affective state from positive to negative, as opposed to being able to flip the robot's "emotional" state like a switch.

The affective state of the robot can also be used to help disambiguate the intent behind utterances with very similar prosodic contours. A good example of this is the difference between utterances intended to soothe versus utterances intended to encourage. The prosodic patterns of these vocalizations are quite similar, but the intent varies with the social context. The communicative function of soothing vocalizations is to comfort a distressed robot— there is no point in comforting the robot if it is not in a distressed state. Hence, the affective assessment phase somatically tags these types of utterances as soothing when the robot is distressed, and as encouraging otherwise (slightly arousing, slightly positive).

7.6   Affective Human-Robot Communication

I have shown that the implemented classifier performs well on the primary caregivers' utterances. Essentially, the classifier is trained to recognize the caregivers' different prosodic contours, which are shown to coincide with Fernald's prototypical patterns. In order to extend the use of the affective intent recognizer, I would like to evaluate the following issues:

• Will naive subjects speak to the robot in an exaggerated manner (in the same way as the caregivers)? Will Kismet's infant-like appearance urge the speakers to use motherese?

• If so, will the classifier be able to recognize the utterances, or will it be hindered by variations in individual's style of speaking or language?

• How will the speakers react to Kismet's expressive feedback, and will the cues encourage them to adjust their speech in a way they think that Kismet will understand?

Five female subjects, ranging from 23 to 54 years old, were asked to interact with Kismet in different languages (English, Russian, French, German, and Indonesian). One of the subjects was a caregiver of Kismet, who spoke to the robot in either English or Indonesian for this experiment. Subjects were instructed to express each affective intent (approval, attention, prohibition, and soothing) and signal when they felt that they had communicated it to the robot. It was expected that many neutral utterances would be spoken during the experiment. All sessions were recorded on video for further evaluations. (Note that similar demonstrations to these experiments can be viewed in the first demonstration, "Recognition of Affective Intent in Robot-Directed Speech," on the included CD-ROM.)

Results

A set of 266 utterances were collected from the experiment sessions. Very long and empty utterances (those containing no voiced segments) were not included. An objective observer was asked to label these utterances and to rate them based on the perceived strength of their affective message (except for neutral). As shown in the classification results (see table 7.6), compared to the caregiver test set, the classifier performs almost as well on neutral, and performs decently well on all the *strong* classes, except for soothing and attentional bids. As expected, the performance reduces as the perceived strength of the utterance decreases.

A closer look at the misclassified soothing utterances showed that a high number of utterances were actually soft approvals. The pitch contours contained a rise-fall segment, but the energy level was low. A linear fit on these contours generates a flat slope, resulting in a neutral classification. A few soothing utterances were confused for neutral despite having the down-sweep frequency characteristic because they contained too many words and coarse pitch contours. Attentional bids generated the worst classification performance

Table 7.6
Classification performance on naive speakers. The subjects spoke to the robot directly and received expressive feedback. An objective scorer ranked each utterance as strong, medium, or weak.

Test Set	Strength	Category	Test Size	Apprv.	Attn.	Prohib.	Sooth.	Neutral	Percent Correct
Care-Givers		Approval	84	64	15	0	5	0	76.2
		Attention	77	21	55	0	5	1	74.3
		Prohibition	80	0	1	78	0	1	97.5
		Soothing	68	0	0	0	55	13	80.9
		Neutral	62	3	4	0	3	52	83.9
Naive Subjects	Strong	Approval	18	14	4	0	0	0	72.2
		Attention	20	10	8	1	0	1	40
		Prohibition	23	0	1	20	0	2	86.9
		Soothing	26	0	1	0	16	10	61.5
	Medium	Approval	20	8	6	0	1	5	40
		Attention	24	10	14	0	0	0	58.3
		Prohibition	36	0	5	12	0	18	33.3
		Soothing	16	0	0	0	8	8	50
	Weak	Approval	14	1	3	0	0	10	7.14
		Attention	16	7	7	0	0	2	43.8
		Prohibition	20	0	4	6	0	10	30
		Soothing	4	0	0	0	0	4	0
		Neutral	29	0	1	0	4	24	82.76

for the strong utterances (it performed better than most for the weak utterances). A careful observation of the classification errors revealed that many of the misclassified attentional bids contained the word "kis-met" spoken with a bell-shaped pitch contour. The classifier recognized this as the characteristic rise-fall pitch segment found in approvals. It was also found that many other common words used in attentional bids, such as "hello" (spoken as "hel-lo-o"), also generated a bell-shaped pitch contour. These are obviously very important issues to be resolved in future efforts to improve the system. Based on these findings, several conclusions can be drawn.

First, a high number of utterances are perceived to carry a *strong* affective message, which implies the use of exaggerated prosody during the interaction session (as hoped for). The remaining question is whether the classifier will generalize to the naive speakers' exaggerated prosodic patterns. Except for the two special cases discussed above, the experimental results indicate that the classifier performs very well in recognizing the naive speakers' prosodic contours even though it was trained only on utterances from the primary caregivers. Moreover, the same failure modes occur in the naive speaker test set. No strongly valenced intents were misclassified as those with opposite valence. It is very encouraging to discover that the classifier not only generalizes to perform well on naive speakers (using either English or other languages), but it also makes very few unacceptable misclassifications.

Discussion

Results from these initial studies and other informal observations suggest that people do naturally exaggerate their prosody (characteristic of motherese) when addressing Kismet. People of different genders and ages often comment that they find the robot to be "cute," which encourages this manner of address. Naive subjects appear to enjoy interacting with Kismet and are often impressed at how life-like it behaves. This also promotes natural interactions with the robot, making it easier for them to engage the robot as if it were a very young child or adored pet.

All female subjects spoke to Kismet using exaggerated prosody characteristic of infant-directed speech. It is quite different from the manner in which they spoke with the experimenters. I have informally noticed the same tendency with children (approximately twelve years of age) and adult males. It is not surprising that individual speaking styles vary. Both children and women (especially women with young children or pets) tend to be uninhibited, whereas adult males are often more reserved. For those who are relatively uninhibited, their styles for conveying affective communicative intent vary. However, Fernald's contours hold for the strongest affective statements in all of the languages that were explored in this study. This would account for the reasonable classifier performance on vocalizations belonging to the strongest affective category of each class. As argued previously, this is the desired behavior for using affective speech as an emotion-based saliency marker for training the robot.

For each trial, we recorded the number of utterances spoken, Kismet's cues, the subject's responses and comments, as well as changes in prosody, if any. Recorded events show that subjects in the study made ready use of Kismet's expressive feedback to assess when the robot "understood" them. The robot's expressive repertoire is quite rich, including both facial expressions and shifts in body posture. The subjects varied in their sensitivity to the robot's expressive feedback, but all used facial expression and/or body posture to determine when the utterance had been properly communicated to the robot. All subjects would reiterate their vocalizations with variations about a theme until they observed the appropriate change in facial expression. If the wrong facial expression appeared, they often used strongly exaggerated prosody to correct the "misunderstanding."

Kismet's expression through face and body posture becomes more intense as the activation level of the corresponding `emotion` process increases. For instance, small smiles versus large grins were often used to discern how "happy" the robot was. Small ear perks versus widened eyes with elevated ears and craning the neck forward were often used to discern growing levels of "interest" and "attention." The subjects could discern these intensity differences, and several modulated their speech to influence them. For example, in one trial a subject scolded Kismet, to which it dipped its head. However, the subject continued to prohibit Kismet with a lower and lower voice until Kismet eventually frowned. Only then did the subject stop her prohibitions.

During course of the interaction, several interesting dynamic social phenomena arose. Often these occurred in the context of prohibiting the robot. For instance, several of the subjects reported experiencing a very strong emotional response immediately after "successfully" prohibiting the robot. In these cases, the robot's saddened face and body posture was enough to arouse a strong sense of empathy. The subject would often immediately stop and look to the experimenter with an anguished expression on her face, claiming to feel "terrible" or "guilty." Subjects were often very apologetic throughout their prohibition session. In this "emotional" feedback cycle, the robot's own affective response to the subject's vocalizations evoked a strong and similar emotional response in the subject as well.

Another interesting social dynamic I observed involved *affective mirroring* between robot and human. In this situation, the subject might first issue a medium-strength prohibition to the robot, which causes it to dip its head. The subject responds by lowering her own head and reiterating the prohibition, this time a bit more foreboding. This causes the robot to dip its head even further and look more dejected. The cycle continues to increase in intensity until it bottoms out with both subject and robot having dramatic body postures and facial expressions that mirror the other. This technique was employed to modulate the degree to which the strength of the message was "communicated" to the robot.

7.7 Limitations and Extensions

The ability of naive subjects to interact with Kismet in this affective and dynamic manner suggests that its response rate is acceptable. The timing delays in the system can and should be improved, however. There is about a 500 ms delay from the time speech ends to receiving an output from the classifier. Much of this delay is due to the underlying speech recognition system, where there is a trade-off between shipping out the speech features to the NT machine immediately after a pause in speech, and waiting long enough during that pause to make sure that speech has completed. There is another delay of approximately one second associated with interpreting the classifier in affective terms and feeding it through to an emotional response. The subject will typically issue one to three short utterances during this time (of a consistent affective content). It is interesting that people rarely seem to issue just one short utterance and wait for a response. Instead, they prefer to communicate affective meanings in a sequence of a few closely related utterances ("That's right, Kismet. Very good! Good robot!"). In practice, people do not seem to be bothered by or notice the delay. The majority of delays involve waiting for a sufficiently strong vocalization to be spoken, since only these are recognized by the system.

Given the motivation of being able to use natural speech as a training signal for Kismet, it remains to be seen how the existing system needs to be improved or changed to serve this purpose. Naturally occurring robot-directed speech doesn't come in nicely packaged sound bites. Often there is clipping, multiple prosodic contours of different types in long utterances, and other background noise (doors slamming, people talking, etc.). Again, targeting infant-caregiver interactions helps alleviate these issues, as infant-directed speech is slower, shorter, and more exaggerated. The collection of robot-directed utterances, however, demonstrates a need to address these issues carefully.

The recognizer in its current implementation is specific to female speakers, and it is particularly tuned to women who can use motherese effectively. Granted, not all people will want to use motherese to instruct robots. At this early state of research, however, I am willing to exploit *naturally occurring* simplifications of robot-directed speech to explore human-style socially situated learning scenarios. Given the classifier's strong performance for the caregivers (those who will instruct the robot intensively), and decent performance for other female speakers (especially for prohibition and approval), I am quite encouraged at these early results. Future improvements include either training a male adult model, or making the current model more gender-neutral.

For instructional purposes, the question remains: *How good is good enough?* A performance of 70 to 80 percent of five-way classifiers for recognizing emotional speech is regarded as state of the art. In practice, within an instructional setting, this may be an unacceptable number of misclassifications. As a result, our approach has taken care to minimize the number of "bad" misclassifications. The social context is also exploited to reduce misclassifications further (such as soothing versus neutral). Finally, expressive feedback is provided to the caregivers so they can make sure that the robot properly "understood" their intent. By incorporating expressive feedback, I have already observed some intriguing social dynamics that arise with naive female subjects. I intend to investigate these social dynamics further so that they can be used to advantage in instructional scenarios.

To provide the human instructor with greater precision in issuing vocal feedback, one must look beyond *how* something is said to *what* is said. Since the underlying speech recognition system (running on the Linux machine) is speaker-independent, this will boost recognition performance for both males and females. It is also a fascinating question of how the robot could learn the valence and arousal associated with particular utterances by bootstrapping from the correlation between those phonemic sequences that show particular persistence during each of the four classes of affective intents. Over time, Kismet could associate the utterance "Good robot!" with positive valence, "No, stop that!" with negative valence, "Look at this!" with increased arousal, and "Oh, it's ok," with decreased arousal by grounding it in an affective context and Kismet's emotional system. Developmental psycholinguists posit that human infants learn their first meanings through this kind of affectively-grounded social interaction with caregivers (Stern et al., 1982). Using punctuated words in this manner gives greater precision to the human caregiver's ability to issue reinforcement, thereby improving the quality of instructive feedback to the robot.

7.8   Summary

Human speech provides a natural and intuitive interface both for communicating with humanoid robots as well as for teaching them. We have implemented and demonstrated a fully integrated system whereby a humanoid robot recognizes and affectively responds to praise, prohibition, attention, and comfort in robot-directed speech. These affective intents are well-matched to human-style instruction scenarios since praise, prohibition, and directing the robot's attention to relevant aspects of a task could be intuitively used to train a robot. Communicative efficacy has been tested and demonstrated with the robot's caregivers as well as with naive subjects. I have argued how such an integrated approach lends robustness to the overall classification performance. Importantly, I have discovered some intriguing social dynamics that arise between robot and human when expressive feedback is introduced. This expressive feedback plays an important role in facilitating natural and intuitive human-robot communication.

8 The Motivation System

In general, animals are in constant battle with many different sources of danger. They must make sure that they get enough to eat, that they do not become dehydrated, that they do not overheat or freeze, that they do not fall victim to a predator, and so forth. The animal's behavior is beautifully adapted to survive and reproduce in this hostile environment. Early ethologists used the term *motivation* to broadly refer to the apparent self-direction of an animal's attention and behavior (Tinbergen, 1951; Lorenz, 1973).

8.1 Motivations in Living Systems

In more evolutionary advanced species, the following features appear to become more prominent: the ability to process more complex stimulus patterns in the environment, the simultaneous existence of a multitude of motivational tendencies, a highly flexible behavioral repertoire, and social interaction as the basis of social organization. Within an animal of sufficient complexity, there are multiple motivating factors that contribute to its observed behavior. Modern ethologists, neuroscientists, and comparative psychologists continue to discover the underlying physiological mechanisms, such as internal clocks, hormones, and internal sense organs, that serve to regulate the animal's interaction with the environment and promote its survival. For the purposes of this chapter, I focus on two classes of motivation systems: homeostatic regulation and emotion.

Homeostatic Regulation

To survive, animals must maintain certain critical parameters within a bounded range. For instance, an animal must regulate its temperature, energy level, amount of fluids, etc. Maintaining each critical parameter requires that the animal come into contact with the corresponding satiatory stimulus (shelter, food, water, etc.) at the right time. The process by which these critical parameters are maintained is generally referred to as *homeostatic regulation* (Carver & Scheier, 1998). In a simplified view, each satiatory stimulus can be thought of as an innately specified need. In broad terms, there is a desired fixed point of operation for each parameter and an allowable bounds of operation around that point. As the critical parameter moves away from the desired point of operation, the animal becomes more strongly motivated to behave in ways that will restore that parameter. The physiological mechanisms that serve to regulate these needs, driving the animal into contact with the needed stimulus at the appropriate time, are quite complex and distinct (Gould, 1982; McFarland & Bosser, 1993).

Emotion

Emotions are another important motivation system for complex organisms. They seem to be centrally involved in determining the behavioral reaction to environmental (often social) and internal events of major significance for the needs and goals of a creature (Plutchik, 1991; Izard, 1977). For instance, Frijda (1994a) suggests that positive emotions are elicited by events that satisfy some motive, enhance one's power of survival, or demonstrate the successful exercise of one's capabilities. Positive emotions often signal that activity toward the goal can terminate, or that resources can be freed for other exploits. In contrast, many negative emotions result from painful sensations or threatening situations. Negative emotions motivate actions to set things right or to prevent unpleasant things from occurring.

Several theorists argue that a few select emotions are *basic* or *primary*—they are endowed by evolution because of their proven ability to facilitate adaptive responses to the vast array of demands and opportunities a creature faces in its daily life (Ekman, 1992; Izard, 1993). The emotions of anger, disgust, fear, joy, sorrow, and surprise are often supported as being basic from evolutionary, developmental, and cross-cultural studies (Ekman & Oster, 1982). Each basic emotion is posited to serve a particular function (often biological or social), arising in particular contexts, to prepare and motivate a creature to respond in adaptive ways. They serve as important reinforcers for learning new behavior. In addition, emotions are refined and new emotions are acquired throughout emotional development. Social experience is believed to play an important role in this process (Ekman & Oster, 1982).

Several theorists argue that emotion has evolved as a relevance-detection and response-preparation system. They posit an appraisal system that assesses the perceived antecedent conditions with respect to the organism's well-being, its plans, and its goals (Levenson, 1994; Izard, 1994; Frijda, 1994c; Lazarus, 1994). Scherer (1994) has studied this assessment process in humans and suggests that people affectively appraise events with respect to novelty, intrinsic pleasantness, goal/need significance, coping, and norm/self compatibility. Hence, the level of cognition required for appraisals can vary widely.

These appraisals (along with other factors such as pain, hormone levels, drives, etc.) evoke a particular emotion that recruits response tendencies within multiple systems. These include physiological changes (such as modulating arousal level via the autonomic nervous system), adjustments in subjective experience, elicitation of behavioral response (such as approach, attack, escape, etc.), and displaying expression. The orchestration of these systems represents a generalized solution for coping with the demands of the original antecedent conditions. Plutchik (1991) calls this stabilizing feedback process *behavioral homeostasis*. Through this process, emotions establish a desired relation between the organism and the environment—pulling toward certain stimuli and events and pushing away from others. Much of the relational activity can be social in nature, motivating proximity seeking, social avoidance, chasing off offenders, etc. (Frijda, 1994b).

The expressive characteristics of emotion in voice, face, gesture, and posture serve an important function in communicating emotional state to others. Levenson (1994) argues that this benefits people in two ways: first, by communicating feelings to others, and second, by influencing others' behavior. For instance, the crying of an infant has a powerful mobilizing influence in calling forth nurturing behaviors of adults. Darwin argued that emotive signaling functions were selected for during the course of evolution because of their communicative efficacy. For members of a social species, the outcome of a particular act usually depends partly on the reactions of the significant others in the encounter. As argued by Scherer, the projection of how the others will react to these different possible courses of action largely determines the creature's behavioral choice. The signaling of emotion communicates the creature's evaluative reaction to a stimulus event (or act) and thus narrows the possible range of behavioral intentions that are likely to be inferred by observers.

Overview of the Motivation System

Kismet's motivations establish its nature by defining its "needs" and influencing how and when it acts to satisfy them. The nature of Kismet is to socially engage people and ultimately to learn from them. Kismet's `drive` and `emotion` processes are designed such that the robot is in homeostatic balance, and an alert and mildly positive affective state, when it is interacting well with people and when the interactions are neither overwhelming nor under-stimulating (Breazeal, 1998). This corresponds to an environment that affords high learning potential as the interactions slightly challenge the robot yet also allow Kismet to perform well.

Kismet's motivation system consists of two related subsystems, one which implements `drives` and a second which implements `emotions`. There are several processes in the emotion system that model different arousal states (such as interest, calm, or boredom). These do not correspond to the *basic emotions,* such as the six proposed by Ekman (anger, disgust, fear, joy, sorrow, and surprise). Nonetheless, they have a corresponding expression and a few have an associated behavioral response. For the purposes here, I will treat these arousal states as `emotions` in this system. Each subsystem serves a regulatory function for the robot (albeit in different ways) to maintain the robot's "well-being." Each `drive` is modeled as an idealized *homeostatic regulation process* that maintains a set of critical parameters within a bounded range. There is one `drive` assigned to each parameter. Kismet's `emotions` are idealized models of basic emotions, where each serves a particular function (often social), each arises in a particular context, and each motivates Kismet to respond in an adaptive manner. They tend to operate on shorter, more immediate, and specific circumstances than the `drives` (which operate over longer time scales).

8.2 The Homeostatic Regulation System

Kismet's `drives` serve four purposes. First, they indirectly influence the attention system. Second, they influence behavior selection by preferentially passing activation to some behaviors over others. Third, they influence the affective state by passing activation energy to the `emotion` processes. Since the robot's expressions reflect its affective state, the `drives` indirectly control the affective cues the robot displays to people. Last, they provide a functional context that organizes behavior and perception. This is of particular importance for emotive appraisals.

The design of Kismet's homeostatic regulation subsystem is heavily inspired by ethological views of the analogous process in animals (McFarland & Bosser, 1993). It is, however, a simplified and idealized model of those discovered in living systems. One distinguishing feature of a `drive` is its temporally cyclic behavior. That is, given no stimulation, a `drive` will tend to increase in intensity unless it is satiated. This is analogous to an animal's degree of hunger or level of fatigue, both following a cyclical pattern.

Another distinguishing feature is its homeostatic nature. Each acts to maintain a level of intensity within a bounded range (neither too much nor too little). Its change in intensity reflects the ongoing needs of the robot and the urgency for tending to them. There is a desired operational point for each `drive` and acceptable bounds of operation around that point. I call this range the *homeostatic regime*. As long as a `drive` is within the homeostatic regime, the robot's needs are being adequately met. For Kismet, maintaining its `drives` within their homeostatic regime is a never-ending process. At any point in time, the robot's behavior is organized about satiating one of its `drives`.

Each `drive` is modeled as a separate process, shown in figure 8.1. Each has a temporal input to implement its cyclic behavior. The activation energy A_{drive} of each `drive` ranges between $[A_{drive}^{-max}, A_{drive}^{+max}]$, where the magnitude of the A_{drive} represents its intensity.

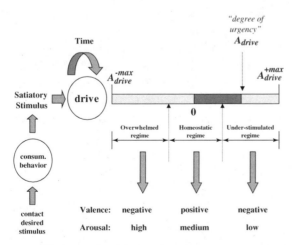

Figure 8.1
The homeostatic model of a `drive` process.

For a given A_{drive} intensity, a large positive magnitude corresponds to under-stimulation by the environment, whereas a large negative magnitude corresponds to over-stimulation by the environment. In general, each A_{drive} is partitioned into three regimes: an *under-stimulated regime,* an *overwhelmed regime,* and the homeostatic regime. A `drive` remains in its homeostatic regime when it is encountering its satiatory stimulus and that stimulus is of appropriate intensity. In the absence of the satiatory stimulus (or if the intensity is too low), the `drive` tends toward the under-stimulated regime. Alternatively, if the satiatory stimulus is too intense, the `drive` tends toward the overwhelmed regime. To remain in balance, it is not sufficient that the satiatory stimulus be present; it must also be of a good quality.

In the current implementation there are three `drives`. They are:

• *Social*

• *Stimulation*

• *Fatigue*

The social drive   The `social-drive` motivates the robot to be in the presence of people and to be stimulated by people. This is important for biasing the robot to learn in a social context. On the under-stimulated extreme, the robot is "lonely"; it is predisposed to act in ways to establish face-to-face contact with people. If left unsatiated, this `drive` will continue to intensify toward the under-stimulated end of the spectrum. On the overwhelmed extreme, the robot is "asocial"; it is predisposed to act in ways to avoid face-to-face contact. The robot tends toward the overwhelmed end of the spectrum when a person is over-stimulating the robot. This may occur when a person is moving too much or is too close to the robot's eyes.

The stimulation drive   The `stimulation-drive` motivates the robot to be stimulated, where the stimulation is generated externally by the environment, typically by engaging the robot with a colorful toy. This `drive` provides Kismet with an innate bias to interact with objects. This encourages the caregiver to draw the robot's attention to toys and events around the robot. On the under-stimulated end of this spectrum, the robot is "bored." This occurs if Kismet has been unstimulated over a period of time. On the overwhelmed part of the spectrum, the robot is "over-stimulated." This occurs when the robot receives more stimulation than its perceptual processes can handle well. In this case, the robot is biased to reduce its interaction with the environment, perhaps by closing its eyes or turning its head away from the stimulus. This `drive` is important for social learning as it encourages the caregiver to challenge the robot with new interactions.

The fatigue drive   The `fatigue-drive` is unlike the others in that its purpose is to allow the robot to shut out the external world instead of trying to regulate its interaction with it. While the robot is "awake," it receives repeated stimulation from the environment or from itself. As time passes, this `drive` approaches the "exhausted" end of the spectrum. Once the intensity level exceeds a certain threshold, it is time for the robot to "sleep." While the robot sleeps, *all* `drives` return to their homeostatic regimes. After this, the robot awakens.

Drives and Affect

The `drives` spread activation energy to the `emotion` processes. In this manner, the robot's ability to satisfy its `drives` and remain in a state of "well-being" is reflected by its affective state. When in the homeostatic regime, a `drive` spreads activation to those processes characterized by positive valence and balanced arousal. This corresponds to a "contented" affective state. When in the under-stimulated regime, a `drive` spreads activation to those processes characterized by negative valence and low arousal. This corresponds to a "bored" affective state that can eventually build to "sorrow." When in the overwhelmed regime, a `drive` spreads activation to those processes characterized by negative valence and high arousal. This corresponds to an affective state of "distress."

The emotion system influences the robot's facial expression. The caregiver can read the robot's facial expression to interpret whether the robot is "distressed" or "content," and can adjust his/her interactions with the robot accordingly. The caregiver accomplishes this by adjusting either the type (social versus non-social) and/or the quality (low intensity, moderate intensity, or high intensity) of the stimulus presented to Kismet. These emotive cues are critical for helping the human work with the robot to establish and maintain a suitable interaction where the robot's `drives` are satisfied, where it is sufficiently challenged, yet where it is largely competent in the exchange.

In chapter 9, I present a detailed example of how the robot's `drives` influence behavior arbitration. In this way, the `drives` motivate which behavior the robot performs to bring itself into contact with needed stimuli.

8.3   The Emotion System

The organization and operation of the emotion system is strongly inspired by various theories of emotions in humans. It is designed to be a flexible system that mediates between both environmental and internal stimulation to elicit an adaptive behavioral response that serves either social or self-maintenance functions (Breazeal, 2001a). The `emotions` are triggered by various events that are evaluated as being of significance to the "well-being" of the robot. Once triggered, each `emotion` serves a particular set of functions to establish a desired relation between the robot and its environment. They motivate the robot to come into contact with things that promote its "well-being" and to avoid those that do not.

Table 8.1
Summary of the antecedents and behavioral responses that comprise Kismet's emotive responses. The antecedents refer to the eliciting perceptual conditions for each `emotion`. The behavior coloumn denotes the observable response that becomes active with the `emotion`. For some, this is simply a facial expression. For others, it is a behavior such as `escape`. The column to the right describes the function each emotive response serves for Kismet.

Antecedent Conditions	Emotion	Behavior	Function
Delay, difficulty in achieving goal of adaptive behavior	anger, frustration	display-displeasure	Show displeasure to caregiver to modify his/her behavior
Presence of an undesired stimulus	disgust	withdraw	Signal rejection of presented stimulus to caregiver
Presence of a threatening, overwhelming stimulus	fear, distress	escape	Move away from a potentially dangerous stimuli
Prolonged presence of a desired stimulus	calm	engage	Continued interaction with a desired stimulus
Success in achieving goal of active behavior, or praise	joy	display-pleasure	Reallocate resources to the next relevant behavior (eventually to reinforce behavior)
Prolonged absence of a desired stimulus, or prohibition	sorrow	display-sorrow	Evoke sympathy and attention from caregiver (eventually to discourage behavior)
A sudden, close stimulus	suprise	startle	Alert
Appearance of a desired stimulus	interest	orient	Attend to new, salient object
Need of an absent and desired stimulus	boredom	seek	Explore environment for desired stimulus

Emotive Responses

This section begins with a high-level discussion of the emotional responses implemented in Kismet. Table 8.1 summarizes under what conditions certain `emotions` and behavioral responses arise, and what function they serve the robot. This table is derived from the evolutionary, cross-species, and social functions hypothesized by Plutchik (1991), Darwin (1872), and Izard (1977). The table includes the six primary emotions proposed by Ekman (i.e., `anger`, `disgust`, `fear`, `joy`, `sorrow`, `surprise`) along with three arousal states (i.e., `boredom`, `interest`, and `calm`). Kismet's expressions of these `emotions` also can be seen on the included CD-ROM in the "Readable Expressions" demonstration.

By adapting these ideas to Kismet, the robot's emotional responses mirror those of biological systems and therefore should seem plausible to a human (please refer to the seventh CD-ROM demonstration titled "Emotive Responses"). This is very important for social interaction. Under close inspection, also note that the four categories of proto-social responses from chapter 3 (affective, exploratory, protective, and regulatory) are represented within this table. Each of the entries in this table has a corresponding affective display. For instance, the robot exhibits sadness upon the prolonged absence of a desired stimulus. This may occur if the robot has not been engaged with a toy for a long time. The sorrowful expression is intended to elicit attentive acts from the human caregiver. Another class of affective responses relates to behavioral performance. For instance, a successfully accomplished goal is reflected by a smile on the robot's face, whereas delayed progress is reflected by a stern expression. Exploratory responses include visual search for desired stimulus and/or maintaining visual engagement of a desired stimulus. Kismet currently has several protective responses, the strongest of which is to close its eyes and turn away from "threatening" or overwhelming stimuli. Many of these emotive responses serve a regulatory function. They bias the robot's behavior to bring it into contact with desired stimuli (orientation or exploration), or to avoid poor quality or "dangerous" stimuli (protection or rejection). In addition, the expression on the robot's face is a social signal to the human caregiver, who responds in a way to further promote the robot's "well-being." Taken as a whole, these affective responses encourage the human to treat Kismet as a socially aware creature and to establish meaningful communication with it.

Components of Emotion

Several theories posit that emotional reactions consist of several distinct but interrelated facets (Scherer, 1984; Izard, 1977). In addition, several appraisal theories hypothesize that a characteristic appraisal (or meaning analysis) triggers the emotional reaction in a context-sensitive manner (Frijda, 1994b; Lazarus, 1994; Scherer, 1994). Summarizing these ideas, an "emotional" reaction for Kismet consists of:

- A precipitating event
- An affective appraisal of that event
- A characteristic expression (face, voice, posture)
- Action tendencies that motivate a behavioral response

Two factors that are not directly addressed with Kismet are:

- Subjective feeling state
- A pattern of physiological activity

Kismet is not conscious, so it does not have feelings.[1] Nor does it have internal sensors that might sense something akin to physiological changes due to autonomic nervous activity. Kismet does, however, have a parameter that maps to arousal level, so in a very simple fashion Kismet has a correlate to autonomic nervous system activity.

1. Several emotion theorists posit that consciousness is a requirement for an organism to experience feeling (see Damasio, 1999). That Kismet is not conscious (at least not yet) is the author's philosophical position.

In living systems, it is believed that these individual facets are organized in a highly interdependent fashion. Physiological activity is hypothesized to physically prepare the creature to act in ways motivated by action tendencies. Furthermore, both the physiological activities and the action tendencies are organized around the adaptive implications of the appraisals that elicited the emotions. From a functional perspective, Smith (1989) and Russell (1997) suggest that the individual components of emotive facial expression are also linked to these emotional facets in a highly systematic fashion.

In the remainder of this chapter, I discuss the relation between the eliciting condition(s), appraisal, action tendency, behavioral response, and observable expression in Kismet's implementation. An overview of the system is shown in figure 8.2. Some of these aspects are covered in greater depth in other chapters. For instance, detailed presentations of the

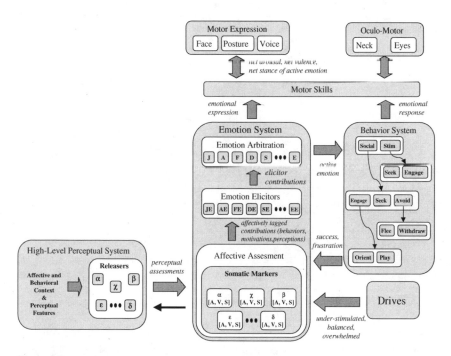

Figure 8.2
An overview of the emotion system. The antecedent conditions come through the high-level perceptual system where they are assessed with respect to the robot's "well-being" and active goals. The result is a set of behavior and emotional response-specific releasers. The emotional response releasers are passed to an affective appraisal phase. In general, behaviors and `drives` can also send influences to this affective appraisal phase. All active contributions are filtered through the emotion elicitors for each `emotion` process. In the emotion arbitration phase, the `emotion` processes compete for activation in a winner-take-all scheme. The winner can evoke its corresponding behavioral response (such as `escape` in the case of `fear`). It also evokes a corresponding facial expression, body posture, and vocal quality. These multi-modality expressive cues are arbitrated by the motor skill system.

expression of affect in Kismet's face, posture, and voice are covered in chapters 10 and 11. A detailed description of how the behavioral responses are implemented is given in chapter 9.

Emotive Releasers

I begin this discussion with the input to the emotion system. The input originates from the high-level perceptual system, where it is fed into an associated *releaser* process. Each releaser can be thought of as a simple "cognitive" assessment that combines lower-level perceptual features into behaviorally significant perceptual categories.

There are many different kinds of releasers defined for Kismet, each hand-crafted, and each combining different contributions from a variety of factors. Each releaser is evaluated with respect to the robot's "well-being" and its goals. This evaluation is converted into an activation level for that releaser. If the perceptual features and evaluation are such that the activation level is above threshold (i.e., the conditions specified by that releaser hold), then its output is passed to its corresponding behavior process in the behavior system. It is also passed to the affective appraisal stage where it can influence the emotion system. There are a number of factors that contribute to the assessment made by each releaser. They are as follows:

• *Drives* The active `drive` provides important context for many releasers. In general, it determines whether a given type of stimulus is either desired or undesired. For instance, if the `social-drive` is active, then skin-toned stimuli are desirable, but colorful stimuli are undesirable (even if they are of good quality). Hence, this motivational context plays an important role in determining whether the emotional response will be one of incorporation or rejection of a presented stimulus.

• *Affective State* The current affective state provides important context for certain releasers. A good example is the `soothing-speech` releaser described in chapter 7. Given a "soothing" classification from the affective intent recognizer, the `soothing-speech` releaser only becomes active if Kismet is "distressed." Otherwise, the `neutral-speech` releaser is activated. This second stage of processing reduces the number of misclassifications between soothing speech versus neutral speech.

• *Active Behavior(s)* The behavioral state also plays an important role in disambiguating certain perceptual conditions. For instance, a `no-face` perceptual condition could correspond to several different possibilities. The robot could be engaged in a `seek-people` behavior, in which case a skin-toned stimulus is a desired but absent stimulus. Initially this would encourage exploration. Over time, however, this could contribute to an state of deprivation due to a long-term loss. Alternatively, the robot could be engaged in an `escape` behavior. In this case, `no-face` corresponds to successful escape, a rewarding circumstance.

• *Perceptual State(s)* The incoming percepts can contribute to the affective state on their own (such as a looming stimulus, for instance), or in combination with other stimuli (such as combining skin-tone with distance to perceive a distant person). An important assessment is how intense the stimulus is. Stimuli that are closer to the robot, move faster, or are larger in the field of view are more intense than stimuli that are further, slower, or smaller. This is an important measure of the quality and threat of the stimulus.

Affective Appraisal

Within the appraisal phase, each releaser with activation above threshold is appraised in affective terms by an associated *somatic marker* (SM) process. Recall from chapter 7 that each active releaser is tagged by affective markers of three types: arousal (A), valence (V), and stance (S). There are four types of appraisals considered:

• *Intensity* The intensity of the stimulus generally maps to arousal. Threatening or very intense stimuli are tagged with high arousal. Absent or low intensity stimuli are tagged with low arousal. Soothing speech has a calming influence on the robot, so it also serves to lower arousal if initially high.

• *Relevance* The relevance of the stimulus (whether it addresses the current goals of the robot) influences valence and stance. Stimuli that are relevant are "desirable" and are tagged with positive valence and approaching stance. Stimuli that are not relevant are "undesirable" and are tagged with negative arousal and withdrawing stance.

• *Intrinsic Pleasantness* Some stimuli are hardwired to influence the robot's affective state in a specific manner. Praising speech is tagged with positive valence and slightly high arousal. Scolding speech is tagged with negative valence and low arousal (tending to elicit `sorrow`). Attentional bids alert the robot and are tagged with medium arousal. Looming stimuli startle the robot and are tagged with high arousal. Threatening stimuli elicit `fear` and are tagged with high arousal, negative valence, and withdrawing stance.

• *Goal Directedness* Each behavior specifies a goal, i.e., a particular relation the robot wants to maintain with the environment. Success in achieving a goal promotes `joy` and is tagged with positive valence. Prolonged delay in achieving a goal results in `frustration` and is tagged with negative valence and withdrawing stance. The stance component increases slowly over time to transition from `frustration` to `anger`.

As initially discussed in chapter 4, because there are potentially many different kinds of factors that modulate the robot's affective state (e.g., behaviors, motivations, perceptions), this tagging process converts the myriad of factors into a common currency that can be combined to determine the net affective state. Further recall that the [A, V, S] trio is the currency the emotion system uses to determine which emotional response should be active.

In the current implementation, the affective tags for each releaser are specified by the designer. These may be fixed constants, or linearly varying quantities. In all, there are three contributing factors to the robot's net affective state:

• *Drives* Recall that each `drive` is partitioned into three regimes: homeostatic, overwhelmed or under-stimulated. For a given `drive`, each regime potentiates arousal and valence differently, which contribute to the activation of different `emotion` processes.

• *Behavior* The success or delayed progress of the active behavior can directly influence the affective state. Success contributes to positive emotive responses, whereas delayed progress contributes to negative emotive responses such as frustration.

• *Releasers* The external environmental factors that elicit emotive responses.

Emotion Elicitors

All somatically marked inputs are passed to the *emotion elicitor* stage. Recall from chapter 7 that the elicitors filter each of the incoming $[A, V, S]$ contributions to determine relevance for its emotive response. Figure 8.3 summarizes how $[A, V, S]$ values map onto each `emotion` process. This filtering is done independently for each type of affective tag. For instance, a valence contribution with a large negative value will not only contribute to the `sad` process, but to the `fear`, `distress`, `anger`, and `disgust` processes as well. Given all these factors, each elicitor computes its average $[A, V, S]$ from all the individual arousal, valence, and stance values that pass through its filter.

Given the net $[A, V, S]$ of an elicitor, the activation level is computed next. Intuitively, the activation level for an elicitor corresponds to how "deeply" the point specified by

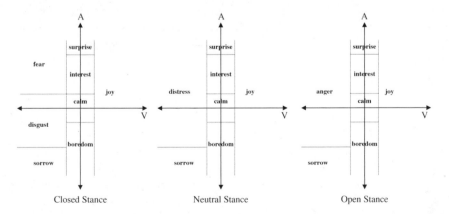

Figure 8.3
Mapping of arousal, valence, and stance dimensions, $[A, V, S]$, to emotions. This figure shows three 2-D slices through this 3-D space.

the net $[A, V, S]$ lies within the arousal, valence, and stance boundaries that define the corresponding `emotion` region shown in figure 8.3. This value is scaled with respect to the size of the region so as to not favor the activation of some processes over others in the arbitration phase. The contribution of each dimension to each elicitor is computed individually. If any one of the dimensions is not represented, then the activation level is set to zero. Otherwise, the A, V, and S contributions are summed together to arrive at the activation level of the elicitor. This activation level is passed on to the corresponding `emotion` process in the arbitration phase.

There are many different processes that contribute to the overall affective state. Influences are sent by `drives`, the active behavior, and releasers. Several different schemes for computing the net contribution to a given `emotion` process were tried, but this one has the nicest properties. In an earlier version, all the incoming contributions were simply averaged. This tended to "smooth" the net affective state to an unacceptable degree. For instance, if the robot's `fatigue-drive` is high (biasing a low arousal state) and a threatening toy appears (contributing to a strong negative valence and high arousal), the averaging technique could result in a slightly negative valence and neutral arousal. This is insufficient to evoke `fear` and an escape response when the robot should protect itself. As an alternative, we could hard-wire certain releasers directly to `emotion` processes. It is not clear, however, how this approach supports the influence of `drives` and behaviors, whose affective contributions change as a function of time. For instance, a given `drive` contributes to `fear`, `sorrow`, or `interest` processes depending on its current activation regime. The current approach balances the constraints of having certain releasers contribute heavily and directly to the appropriate emotive response, while accommodating those influences that contribute to different `emotions` as a function of time. The end result also has nice properties for generating facial expressions that reflect this assessment process in a rich way. This is important for social interaction as originally argued by Darwin. This expressive benefit is discussed in further detail in chapter 10.

Emotion Activation

Next, the activation level of each `emotion` process is computed. There is a process defined for each `emotion` listed in table 8.1: `joy`, `anger`, `disgust`, `fear`, `sorrow`, `surprise`, `interest`, `boredom`, and `calm`.

Numerically, the activation level $A_{emotion}$ of each `emotion` process can range between $[0, A_{emotion}^{max}]$ where $A_{emotion}^{max}$ is an integer value determined empirically. Although these processes are always active, their intensity must exceed a threshold level before they are expressed externally. The activation of each process is computed by the equation:

$$A_{emotion} = \sum (E_{emotion} + B_{emotion} + P_{emotion}) - \delta_t$$

where $E_{emotion}$ is the activation level of its affiliated elicitor process; $B_{emotion}$ is a DC bias that can be used to make some `emotion` processes easier to activate than others. $P_{emotion}$ adds a level of persistence to the active emotion. This introduces a form of inertia so that different `emotion` processes don't rapidly switch back and forth. Finally, δ_t is a decay term that restores an `emotion` to its bias value once the `emotion` becomes active. Hence, unlike `drives` (which contribute to the robot's longer-term "mood"), the `emotions` have an intense expression followed by decay to a baseline intensity. The decay takes place on the order of seconds.

Emotion Arbitration

Next, the `emotion` processes compete for control in a winner-take-all arbitration scheme based on their activation level. The activation level of an `emotion` process is a measure of its relevance to the current situation. Each of these processes is distinct from the others and regulates the robot's interaction with its environment in a distinct manner. Each becomes active in a different environmental (or internal) situation. Each motivates a different observable response by spreading activation to a specific behavior process in the behavior system. If this amount of activation is strong enough, then the active `emotion` can "seize" temporary control and force the behavior to become expressed. In a process of behavioral homeostasis as proposed by Plutchik (1991), the emotive response maintains activity through feedback until the correct relation of robot to environment is established.

Concurrently, the net $[A, V, S]$ of the active process is sent to the expressive components of the motor system, causing a distinct facial expression, vocal quality, and body posture to be exhibited. The strength of the facial expression reflects the level of activation of the `emotion`. Figure 8.4 illustrates the emotional response network for the `fear` process. Affective networks for the other responses in table 8.1 are defined in a similar manner. By modeling Kismet's emotional responses after those of living systems, people have a natural and intuitive understanding of Kismet's "emotional" behavior and how to influence it.

There are two threshold levels for each `emotion` process: one for expression and one for behavioral response. The expression threshold is lower than the behavior threshold. This allows the facial expression to lead the behavioral response. This enhances the readability and interpretation of the robot's behavior for the human observer. For instance, if the caregiver shakes a toy in a threatening manner near the robot's face, Kismet will first exhibit a fearful expression and then activate the escape response. By staging the response in this manner, the caregiver gets immediate expressive feedback that she is "frightening" the robot. If this was not the intent, then the caregiver has an intuitive understanding of why the robot appears frightened and modifies behavior accordingly. The facial expression also sets up the human's expectation of what behavior will soon follow. As a result, the caregiver

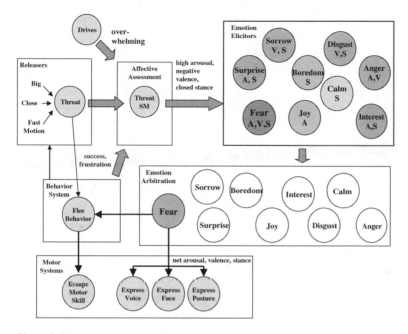

Figure 8.4
The implementation of the `fear` process. The releaser for `threat` is passed to the affective assessment phase. It is tagged with high arousal, negative valence, and closed stance by the corresponding somatic marker process. This affective information is then filtered by the corresponding elicitor of each emotion process. Darker shading corresponds to a higher activation level. Note that only the `fear-elicitor` process has each of the arousal, valence, and stance conditions matched (hence, it has the darkest shading). As a result, it is the only one that passes activation to its corresponding `emotion` process.

not only sees what the robot is doing, but has an understanding of why. (An example of these behaviors can be viewed on the included CD-ROM's "Emotive Responses" section.)

8.4 Regulating Playful Interactions

Kismet's design relies on the ability of people to interpret and understand the robot's behavior. If this is the case, then the robot can use expressive feedback to tune the caregiver's behavior in a manner that benefits the interaction.

In general, when a `drive` is in its homeostatic regime, it potentiates positive valenced `emotions` such as `joy` and arousal states such as `interest`. The accompanying expression tells the human that the interaction is going well and the robot is poised to play (and ultimately learn). When a `drive` is not within the homeostatic regime, negative valenced emotions are potentiated (such as anger, fear, or sorrow), which produces signs of distress on the robot's face. The particular sign of distress provides the human with additional cues as to what is "wrong" and how he/she might correct for it. For example, overwhelming stimuli (such as a rapidly moving toy) produce signs of fear. Similarly, Infants often show signs of anxiety when placed in a confusing environment.

Note that the same sort of interaction can have a very different "emotional" effect on the robot depending on the motivational context. For instance, playing with the robot while all drives are within the homeostatic regime elicits joy. This tells the human that playing with the robot is a good interaction to be having at this time. If, however, the fatigue-drive is deep into the under-stimulated end of the spectrum, then playing with the robot actually prevents the robot from going to "sleep." As a result, the fatigue-drive continues to increase in intensity. When high enough, the fatigue-drive begins to potentiate anger since the goal of sleep is blocked. The human may interpret this as the robot acting cranky because it is "tired."

In this section I present a couple of interaction experiments to illustrate how the robot's motivations and facial expressions can be used to regulate the nature and quality of social exchange with a person. Several chapters in this book give other examples of this process (chapters 7 and 12 in particular). Whereas the examples in this chapter focus on the interaction of emotions, drives, and expression, other chapters focus on the perceptual conditions of eliciting different emotive responses.

Each experiment involves a caregiver interacting with the robot using a colorful toy. Data was recorded on-line in real-time during the exchange. Figures 8.5 and 8.6 plot the activation levels of the appropriate emotions, drives, behaviors, and percepts. Emotions are always plotted together with activation levels ranging from 0 to 2000. Percepts, behaviors, and drives are often plotted together. Percepts and behaviors have activation levels that also range from 0 to 2000, with higher values indicating stronger stimuli or higher potentiation respectively. Drives have activation ranging from -2000 (the overwhelmed extreme) to 2000 (the under-stimulated extreme). The perceptual system classifies the toy as a non-face stimuli, thus it serves to satiate the stimulation drive. The motion generated by the object gives a rating of the stimulus intensity. The robot's facial expressions reflect its ongoing motivational state and provides the human with visual cues as to how to modify the interaction to keep the robot's drives within homeostatic ranges.

For the waving toy experiment, a lack of interaction before the start of the run ($t \leq 0$) places the robot in a sad emotional state as the stimulation-drive lies in the under-stimulated end of the spectrum for activation $A_{stimulation} \geq 400$. This corresponds to a long-term loss of a desired stimulus. From $5 \leq t \leq 25$ a salient toy appears and stimulates the robot within the acceptable intensity range ($400 \leq A_{nonFace} \leq 1600$) on average. This corresponds to waving the toy gently in front of the robot. This amount of stimulus causes the

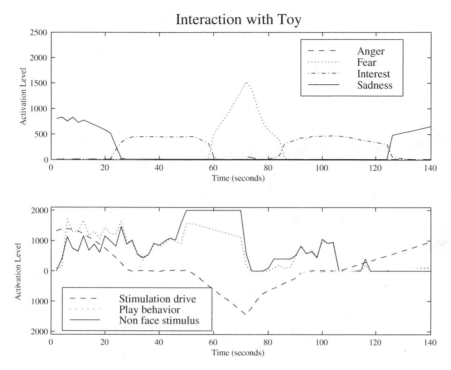

Figure 8.5
Experimental results for the robot interacting with a person waving a toy. The top chart shows the activation levels of the `emotions` involved in this experiment as a function of time. The bottom chart shows the activation levels of the `drives`, behaviors, and percepts relevant to this experiment.

`stimulation-drive` to diminish until it resides within the homeostatic range, and a look of interest appears on the robot's face. From $25 \leq t \leq 45$ the stimulus maintains a desirable intensity level, the `drive` remains in the homeostatic regime, and the robot maintains `interest`.

At $45 \leq t \leq 70$ the toy stimulus intensifies to large, sweeping motions that threaten the robot ($A_{nonFace} \geq 1600$). This causes the `stimulation-drive` to migrate toward the overwhelmed end of the spectrum and the `fear` process to become active. As the `drive` approaches the overwhelmed extreme, the robot's face displays an intensifying expression of fear. Around $t = 75$ the expression peaks at an emotional level of $A_{fear} = 1500$ and experimenter responds by stopping the waving stimulus before the escape response is triggered. With the threat gone, the robot "calms" somewhat as the `fear` process decays. The interaction then resumes at an acceptable intensity. Consequently, the `stimulation-drive` returns to the homeostatic regime and the robot displays interest again. At $t \geq 105$ the

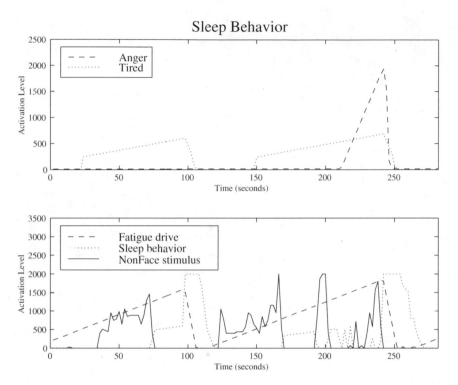

Figure 8.6
Experimental results for long-term interactions of the `fatigue-drive` and the `sleep` behavior. The `fatigue-drive` continues to increase until it reaches an activation level that potentiates the `sleep` behavior. If there is no other stimulation, this will allow the robot to activate the `sleep` behavior.

waving stimulus stops for the remainder of the run. Because of the prolonged loss of the desired stimulus, the robot is under-stimulated and an expression of sadness reappears on the robot's face.

Figure 8.6 illustrates the influence of the `fatigue-drive` on the robot's motivational and behavioral state when interacting with a caregiver. Over time, the `fatigue-drive` increases toward the under-stimulated end of the spectrum. As the robot's level of "fatigue" increases, the robot displays stronger signs of being tired. At time step $t = 95$, the `fatigue-drive` moves above the threshold value of 1600, which is sufficient to activate the `sleep` behavior when no other interactions are occurring. The robot remains "asleep" until all `drives` are restored to their homeostatic ranges. Once this occurs, the activation level of the `sleep` behavior decays until the behavior is no longer active and the robot "wakes up" in an calm state.

At time step $t = 215$, the plot shows what happens if a human continues to interact with the robot despite its "fatigued" state. The robot cannot "fall asleep" as long as the play-with-toy behavior wins the competition and inhibits the sleep behavior. If the fatigue-drive exceeds threshold and the robot cannot fall asleep, the robot begins to show signs of frustration. Eventually the robot's "frustration" increases until the robot achieves anger (at $t = 1800$). Still the human persists with the interaction. Eventually the robot's fatigue-level reaches near maximum, and the sleep behavior wins out.

These experiments illustrate a few of the emotive responses of table 8.1 that arise when engaging a human. It demonstrates how the robot's emotive cues can be used to regulate the nature and intensity of the interaction, and how the nature of the interaction influences the robot's behavior. (Additional video demonstrations can be viewed on the included CD-ROM.) The result is an ongoing "dance" between robot and human aimed at maintaining the robot's drives within homeostatic bounds and maintaining a good affective state. If the robot and human are good partners, the robot remains "interested" most of the time. These expressions indicate that the interaction is of appropriate intensity for the robot.

8.5   Limitations and Extensions

Kismet's motivation system appears adequate for generating infant-like social exchanges with a human caregiver. To incorporate social learning, or to explore socio-emotional development, a number of extensions could be made.

Extension to drives   To support social learning, new drives could be incorporated into the system. For instance, a *self-stimulation drive* could motivate the robot to play by itself, perhaps modulating its vocalizations to learn how to control its voice to achieve specific auditory effects. A *mastery/curiosity drive* might motivate the robot to balance exploration versus exploitation when learning new skills. This would correlate to the amount of novelty the robot experiences over time. If its environment is too predictable, this drive could bias the robot to prefer novel situations. If the environment is highly unpredictable for the robot, it could show distress, which would encourage the caregiver to slow down.

Ultimately, the drives should provide the robot with a reinforcement signal as Blumberg (1996) has done. This could be used to motivate the robot to learn communication skills that satisfy its drives. For instance, the robot may discover that making a particular vocalization results in having a toy appear. This has the additional effect that the stimulation-drive becomes satiated. Over time, through repeated games with the caregiver, the caregiver could treat that particular vocalization as a request for a specific toy. Given enough of these consistent, contingent interactions during play, the robot may learn to utter that vocalization with the *expectation* that its `stimulation-drive` be reduced. This would constitute a simple act of meaning.

Extensions to emotions   Kismet's `drives` relate to a hardwired preference for certain kinds of stimuli. The power of the emotion system is its ability to associate affective qualities to different kinds of events and stimuli. As discussed in chapter 7, the robot could have a learning mechanism by which it uses the caregiver's affective assessment (praise or prohibition) to affectively tag a particular object or action. This is of particular importance if the robot is to learn something novel—i.e., something for which it does not already have an explicit evaluation function. Through a process of social referencing (discussed in chapter 3) the robot could learn how to organize its behavior using the caregiver's affective assessment. Human infants continually encounter novel situations, and social referencing plays an important role in their cognitive, behavioral, and social development.

Another aspect of learning involves learning new `emotions`. These are termed *secondary emotions* (Damasio, 1994). Many of these are socially constructed through interactions with others.

As done in Picard (1997), one might pose the question, "What would it take to give Kismet genuine emotions?" Kismet's emotion system addresses some of the aspects of emotions in simple ways. For instance, the robot carries out some simple "cognitive" appraisals. The robot expresses its "emotional" state. It also uses analogs of emotive responses to regulate its interaction with the environment to promote its "well-being." There are many aspects of human emotions that the system does not address, however, nor does it address any at an adult human level.

For instance, many of the appraisals proposed by (Scherer, 1994) are highly cognitive and require substantial social knowledge and self awareness. The robot does not have any "feeling" states. It is unclear if consciousness is required for this, or what consciousness would even mean for a robot. Kismet does not reason about the emotional state of others. There have been a few systems that have been designed for this competence that employ symbolic models (Ortony et al., 1988; Elliot, 1992; Reilly, 1996). The ability to recognize, understand, and reason about another's emotional state is an important ability for having a theory of mind about other people, which is considered by many to be a requisite of adult-level social intelligence (Dennett, 1987).

Another aspect I have not addressed is the relation between emotional behavior and personality. Some systems tune the parameters of their emotion systems to produce synthetic characters with different personalities—for instance, characters who are quick to anger, more timid, friendly, and so forth (Yoon et al., 2000). In a similar manner, Kismet has its own version of a synthetic personality, but I have tuned it to this particular robot and have not tried to experiment with different synthetic personalities. This could be an interesting set of studies.

This leads us to a discussion of both an important feature and limitation of the motivation system—the number of parameters. Motivation systems of this nature are capable of producing rich, dynamic, compelling behavior at the expense of having many parameters that must be tuned. For this reason, systems of the complexity that rival Kismet are hand-crafted. If learning is introduced, it is done so in limited ways. This is a trade-off of the technique, and there are no obvious solutions. Designers scale the complexity of these systems by maintaining a principled way of introducing new releasers, appraisals, elicitors, etc. The functional boundaries and interfaces between these stages must be honored.

8.6   Summary

Kismet's emotive responses enable the robot to use social cues to tune the caregiver's behavior so that both perform well during the interaction. Kismet's motivation system is explicitly designed so that a state of "well-being" for the robot corresponds to an environment that affords a high learning potential. This often maps to having a caregiver actively engaging the robot in a manner that is neither under-stimulating nor overwhelming. Furthermore, the robot actively regulates the relation between itself and its environment, to bring itself into contact with desired stimuli and to avoid undesired stimuli. All the while, the cognitive appraisals leading to these actions are displayed on the robot's face. Taken as a whole, the observable behavior that results from these mechanisms conveys intentionality to the observer. This is not surprising as they are well-matched to the proto-social responses of human infants. In numerous examples presented throughout this book, people interpret Kismet's behavior as the product of intents, beliefs, desires, and feelings. They respond to Kismet's behaviors in these terms. This produces natural and intuitive social exchange on a physical and affective level.

9 The Behavior System

With respect to social interaction, Kismet's behavior system must be able to support the kinds of behaviors that infants engage in. Furthermore, it should be initially configured to emulate those key action patterns observed in an infant's initial repertoire that allow him/her to interact socially with the caregiver. Because the infant's initial responses are often described in ethological terms, the architecture of the behavior system adopts several key concepts from ethology regarding the organization of behavior (Tinbergen, 1951; Lorenz, 1973; McFarland & Bosser, 1993; Gould, 1982).

Several key action patterns that serve to foster social interaction between infants and their caregivers can be extracted from the literature on pre-speech communication of infants (Bullowa, 1979; de Boysson-Bardies, 1999). In chapter 3, I discussed these action patterns, the role they play in establishing social exchanges with the caregiver, and the importance of these exchanges for learning meaningful communication acts. Chapter 8 presented how the robot's homeostatic regulation mechanisms and emotional models take part in many of these proto-social responses. This chapter presents the contributions of the behavior system to these responses.

9.1 Infant-Caregiver Interaction

Tronick et al. (1979) identify five phases that characterize social exchanges between three-month-old infants and their caregivers: initiation, mutual-orientation, greeting, play-dialogue and disengagement. As introduced in chapter 3, each phase represents a collection of behaviors that mark the state of the communication. Not every phase is present in every interaction, and a sequence of phases may appear multiple times within a given exchange, such as repeated greetings before the play-dialogue phase begins, or cycles of disengagement to mutual orientation to disengagement. Hence, the order in which these phases appear is somewhat flexible yet there is a recognizable structure to the pattern of interaction. These phases are described below:

• *Initiation* In this phase, one of the partners is involved but the other is not. Frequently it is the mother who tries to actively engage her infant. She typically moves her face into an in-line position, modulates her voice in a manner characteristic of attentional bids, and generally tries to get the infant to orient toward her. Chapters 6 and 7 present how these cues are naturally and intuitively used by naive subjects to get Kismet's attention.

• *Mutual Orientation* Here, both partners attend to the other. Their faces may be either neutral or bright. The mother often smoothes her manner of speech, and the infant may make isolated sounds. Kismet's ability to locate eyes in its visual field and direct its gaze toward them is particularly powerful during this phase.

• *Greeting* Both partners attend to the other as smiles are exchanged. Often, when the baby smiles, his limbs go into motion and the mother becomes increasingly animated. (This is the case for Kismet's greeting response where the robot's smile is accompanied by small ear motions.) Afterwards, the infant and caregiver move to neutral or bright faces. Now they may transition back to mutual orientation, initiate another greeting, enter into a play dialogue, or disengage.

• *Play Dialogue* During this phase, the mother speaks in a burst-pause pattern and the infant vocalizes during the pauses (or makes movements of intention to do so). The mother responds with a change in facial expression or a single burst of vocalization. In general, this phase is characterized by mutual positive affect conveyed by both partners. Over time the affective level decreases and the infant looks away.

• *Disengagement* Finally, one of the partners looks away while the other is still oriented. Both may then disengage, or one may try to reinitiate the exchange.

Proto-Social Skills for Kismet

In chapter 3, I categorized a variety of infant proto-social responses into four categories (Breazeal & Scassellati, 1999b). With respect to Kismet, the affective responses are important because they allow the caregiver to attribute feelings to the robot, which encourages the human to modify the interaction to bring Kismet into a positive emotional state. The exploratory responses are important because they allow the caregiver to attribute curiosity, interest, and desires to the robot. The human can use these responses to direct the interaction toward things and events in the world. The protective responses are important to keep the robot from damaging stimuli, but also to elicit concern and caring responses from the caregiver. The regulatory responses are important for pacing the interaction at a level that is suitable for both human and robot.

In addition, Kismet needs skills that allow it to engage the caregiver in tightly coupled dynamic interactions. Turn-taking is one such skill that is critical to this process (Garvey, 1974). It enables the robot to respond to the human's attempts at communication in a tightly temporally correlated and contingent manner. If the communication modality is facial expression, then the interaction may take the form of an imitative game (Eckerman & Stein, 1987). If the modality is vocal, then proto-dialogues can be established (Rutter & Durkin, 1987; Breazeal, 2000b). This dynamic is a cornerstone of the social learning process that transpires between infant and adult.

9.2   Lessons from Ethology

For Kismet to engage a human in this dynamic, natural, and flexible manner, its behavior needs to be robust, responsive, appropriate, coherent, and directed. Much can be learned from the behavior of animals, who must behave effectively in a complex dynamic environment in order to satisfy their needs and maintain their well-being. This entails having the animal apply its limited resources (finite number of sensors, muscles and limbs, energy, etc.) to perform numerous tasks. Given a specific task, the animal exhibits a reasonable amount of persistence. It works to accomplish a goal, but not at the risk of ignoring other important tasks if the current task is taking too long.

For ethologists, the animal's observable behavior attempts to satisfy its competing physiological needs in an uncertain environment. Animals have multiple needs that must be tended to, but typically only one need can be satisfied at a time (hunger, thirst, rest, etc.). Ethologists strive to understand how animals organize their behaviors and arbitrate between them to satisfy these competing goals, how animals decide what to do for how long, and how they decide which opportunities to exploit (Gallistel, 1980).

By observing animals in their natural environment, ethologists have made significant contributions to understanding animal behavior and providing descriptive models to explain its organization and characteristics. In this section, I present several key ideas from ethology that have strongly influenced the design of the behavior system. These theories and concepts specifically address the issues of relevance, coherence, and concurrency, which are critical for animal behavior as well as for the robot's behavior. The behavior system I have constructed is similar in spirit to that of Blumberg (1996), who has also drawn significant insights from animal behavior.

Behaviors

Ethologists such as Lorenz (1973) and Tinbergen (1951) viewed behaviors as being complex, temporally extended patterns of activity that address a specific biological need. In general, the animal can only pursue one behavior at a time such as feeding, defending territory, or sleeping. As such, each behavior is viewed as a self-interested goal-directed entity that competes against other behaviors for control of the creature. They compete for expression based on a measure of relevance to the current internal and external situation. Each behavior determines its own degree of relevance by taking into account the creature's internal motivational state and its perceived environment.

Perceptual Contributions

For the perceptual contribution to behavioral relevance, Tinbergen and Lorenz posited the existence of innate and highly schematic perceptual filters called *releasers*. Each releaser is an abstraction for the minimal collection of perceptual features that reliably identify a particular object or event of biological significance in the animal's natural environment. Each releaser serves as the perceptual elicitor to either a group of behaviors or to a single behavior. The function of each releaser is to determine if all perceptual conditions are right for its affiliated behavior to become active. Because each releaser is not overly specific or precise, it is possible to "fool" the animal by devising a mock stimulus that has the right combination of features to elicit the behavioral response. In general, releasers are conceptualized to be simple, fast, and just adequate. When engaged in a particular behavior, the animal tends to only attend to those features that characterize its releaser.

Motivational Contributions

Ethologists have long recognized that an animal's internal factors contribute to behavioral relevance. I discussed two examples of motivating factors in chapter 8, namely homeostatic regulatory mechanisms and emotions. Both serve regulatory functions for the animal to maintain its state of well-being. The homeostatic mechanisms often work on slower time-scales and bring the animal into contact with innately specified needs, such as food, shelter, and water. The emotions operate on faster time-scales and regulate the relation of the animal with its (often social) environment. An active emotional response can be thought of as temporarily seizing control of the behavior system to force the activation of a particular observable response in the absence of other contributing factors. By doing so, the emotion addresses the antecedent conditions that evoked it. Emotions bring the animal close to things that benefit its survival, and motivate it to avoid those circumstances that are detrimental to its well-being. Emotional responses are also highly adaptive, and the animal can learn how to apply them to new circumstances.

Overall, motivations add richness and complexity to an animal's behavior, far beyond a stimulus-response or reflexive sort of behavior that might occur if only perceptual inputs were considered, or if there were a simple hardwired mapping. Motivations determine the internal agenda of the animal, which changes over time. As a result, the same perceptual stimulus may result in a very different behavior. Or conversely, very different perceptual stimuli may result in an identical behavior given a different motivational state. The motivational state will also affect the strength of perceptual stimuli required to trigger a behavior. If the motivations heavily predispose a particular behavior to be active, a weak stimulus might be sufficient to activate the behavior. Conversely, if the motivations contribute minimally, a very strong stimulus is required to activate the behavior. Scherer (1994a) discusses the advantages of having emotions decouple the stimulus from the response in emotive reactions. For members in a social species, one advantage is the latency this decoupling introduces between affective expression and ensuing behavioral response. This makes an animal's behavior more readable and predictable to the other animals that are in close contact.

Behavior Groups

Up to this point, I have taken a rather simplified view of behavior. In reality, a behavior to reduce hunger may be composed of collections of related behaviors. Within each group, behaviors are activated in turn, which produces a sequence of distinguishable motor acts. For instance, one behavior may be responsible for eating while the others are responsible for bringing the animal near food. In this case, eating is the *consummatory behavior* because it serves to directly satiate the affiliated hunger drive when active. It is the last behavior activated in a sequence simply because once the drive is satiated, the motivation for engaging in the eating behavior is no longer present. This frees the animal's resources to tend to other needs. The other behaviors in the group are called *appetitive behaviors*. The appetitive behaviors represent separate behavioral strategies for bringing the animal to a relationship with its environment where it can directly activate the desired consummatory behavior. Lorenz considered the consummatory behavior to constitute the "goal" of the preceding appetitive behaviors. The appetitive behaviors "seek out" the appropriate releaser that will ultimately result in the desired consummatory behavior.

Given that each behavior group is composed of competing behaviors, a mechanism is needed to arbitrate between them. For appropriately persistent behavior, the arbitration mechanism should have some "inertia" term which allows the currently active behavior enough time to achieve its goal. If the active behavior's rate of progress is too slow, however, it should eventually allow other behaviors to become active. Some behaviors (such as feeding) might have a higher priority than other behaviors (such as preening), yet sometimes it is important for the preening behavior to be preferentially activated. Hence, the creature must perform "time-sharing," where lower priority activities are given a chance to execute despite the presence of a higher priority activity.

Behavior Hierarchies

Tinbergen's *hierarchy of behavior centers* (an example is shown in figure 9.1) is a more general explanation of behavioral choice that incorporates many of the ideas mentioned above (Tinbergen, 1951). It accounts for behavioral sequences that link appetitive behaviors to the desired consummatory behavior. It also factors in both perceptual and internal factors in behavior selection.

In Tinbergen's hierarchy, the nodes stand for behavior centers and the links symbolize transfer of energy between nodes. Behaviors are categorized according to function (i.e., which biological need it serves). Each class of behavior is given a separate hierarchy. For instance, behaviors such as feeding, defending territory, procreation, etc., are placed at the pinnacle of their respective hierarchies. These top-level centers must be "motivated" by a form of energy—i.e., drive factors. Figure 9.1 is Tinbergen's proposed model to explain the procreating behavior of the male stickleback fish.

Activation energy is specific to an entire category of behavior (its respective hierarchy) and can "flow" down the hierarchy to motivate the behavior centers (groups of behaviors). Paths from the top-level center pass the energy to subordinate centers, but only if the correct

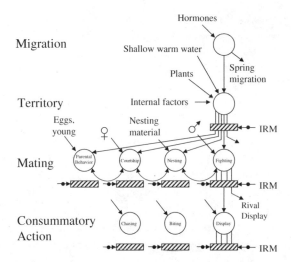

Figure 9.1
Tinbergen's proposed hierarchy to model the procreation behavior of the male stickleback fish (adapted from Tinbergen [1951]). The motivational influences (hormones, etc.) operate at the top level. Behaviors of increasing specificity are modeled at deeper levels in the hierarchy. The motor responses are at the bottom.

perceptual conditions for that behavior center are present. Such percept-based blocks are represented as rectangles under each node in figure 9.1. Until the appropriate stimulus is encountered, a behavior center under the block will not be executed. When stimulated, the block is removed and the flow of energy allows the behaviors within the group to execute and subsequently to pass activation to lower centers.

The hierarchical structure of behavior centers ensures that the creature will perform the sort of activity that will bring it face-to-face with the appropriate stimulus to release the lower level of behavior. Downward flow of energy allows appetitive behaviors to be activated in the correct sequence. Several computational models of behavior selection have used a similar mechanism, such as in Tyrrell (1994) and Blumberg (1994). Implicit in this model is that at every level of the hierarchy, a "decision" is being made among several alternatives, of which one is chosen. At the top, the decisions are very general (feed versus drink) and become increasingly more specific as one moves down a hierarchy.

9.3   Organization of Kismet's Behavior System

Following an ethological perspective and previously noted works, Kismet's behavior system organizes the robot's goals into a coherent structure (see figure 9.2). Each behavior is viewed as a self-interested, goal-directed entity that competes with other behaviors to establish the

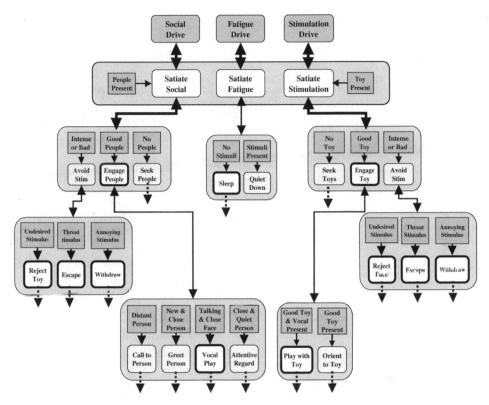

Figure 9.2
Kismet's behavior hierarchy. Bold nodes correspond to consummatory behavior(s) of the behavior group. Solid lines pass activation to other behaviors. Dashed lines send requests to the motor system. The emotional influences are not shown at this scale.

current task of the robot. Given that the robot has multiple time-varying goals that it must tend to, and different behavioral strategies that it can employ to achieve them, an arbitration mechanism is required to determine which behavior(s) to activate and for how long. The main responsibility of the behavior system is to carry out this arbitration. By doing so, it addresses the issues of relevancy, coherency, concurrency, persistence, and opportunism as discussed in chapter 4. Note, that to perform the behavior, the behavior system must work in concert with the motor systems (see chapters 10, 11, and 12). The motor systems are responsible for controlling the robot's motor modalities such that the stated goal of the behavior system is achieved.

The behavior system is organized into loosely layered, heterogeneous hierarchies of behavior groups (Blumberg, 1994). Each group contains behaviors that compete for activation with one another. At the highest level, behaviors are organized into competing *functional groups* (the primary branches of the hierarchy) where each group is responsible for maintaining one of the three homeostatic functions (i.e., to be social, to be stimulated by the environment, and to occasionally rest).

Only one functional group can be active at a time. The influence of the robot's `drives` is strongest at the top level of the hierarchy, biasing which functional group should be active. This motivates the robot to come into contact with the satiatory stimulus for that `drive`. The intensity level of the `drive` being tended to biases behavior to establish homeostatic balance. This is described in more detail in section 9.4.

The "emotional" influence on behavior activation is more direct and immediate. As discussed in chapter 8, each emotional response is mapped to a distinct behavioral response. Instead of influencing behavior only at the top level of the hierarchy (as is the case with `drives`), an active `emotion` directly activates the coordinating behavioral response. It accomplishes this by sending sufficient activation energy to its affiliated behavior(s) and behavior groups such that the desired behavior wins the competition among other behaviors and becomes active. In this way, an `emotion` can "hijack" behavior to suit its own purposes.

Each functional group consists of an organized hierarchy of behavior groups. At each level in the hierarchy, each behavior group represents a competing strategy (a collection of behaviors) for satisfying the goal of its parent behavior. In turn, each behavior within a behavior group is viewed as a task-achieving entity whose particular goal contributes to the strategy of its behavior group. The behavior groups are akin to Tinbergen's *behavioral centers*. They are represented as *container nodes* in the hierarchy (because they "contain" the competing behaviors of that group). They are similar in spirit to the behavior groups of Blumberg's system, however, whereas Blumberg (1994) uses mutual inhibition between competing behaviors within a group to determine the winner, the container node compares the activation levels of its behaviors to determine the winner.

Each behavior group consists of a consummatory behavior and one or more appetitive behaviors. The goal of a behavior group is to activate the consummatory behavior of that group. When the consummatory behavior is carried out, the task of that behavior group is achieved. Each appetitive behavior is designed to bring the robot into a relationship with the environment so that its associated consummatory behavior is activated. A given appetitive behavior might require the performance of other more specific tasks. In this case, these more specific tasks are represented as a child behavior group of the appetitive behavior. Each child behavior group represents a different strategy for achieving the parent (Blumberg, 1996).

Hence, at the behavioral category level, the functional groups compete to determine which need is to be met (socializing, playing, or sleeping). At the strategy level, behavior groups of the winning functional group compete for expression. Finally, on the task level, the behaviors of the winning behavior group compete for expression. As with Blumberg's

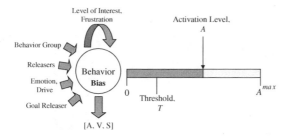

Level of Interest,
Frustration

Activation Level,
A

Behavior Group

Releasers

Emotion,
Drive

Goal Releaser

Behavior
Bias

0

Threshold,
T

A^{max}

[A, V, S]

Figure 9.3
The model of a behavior.

system, the observed behavior of the robot is the result of competition at the functional, strategy, and task levels.

The Behavior Model

The individual behaviors within a group compete for activation based on their computed relevance to the given situation. Each behavior determines its own relevance by taking into account perceptual factors (as defined by its affiliated releaser and goal releaser) as well as internal factors (see figure 9.3). The internal factors can either arise from an affiliated `emotion` (or `drive` at the top level), from activity of the behavior group to which it belongs (or the child behavior group, if present), or the behavior's own internal state (such as its frustration, current level of interest, or prepotentiated bias). Hence, as was the case with the motivational system, there are many different types of factors that contribute to a behavior's relevance. These influences must be converted into a common currency and combined to compute the activation level for the behavior. The activation level represents some measure of the behavior's "value" to the robot at that point in time.

Provided that the behavior group is active, each behavior within the group updates its level of activation by the equation:

$$A_{update} = \sum_{n}(releaser_n \cdot gain_n) + \sum_{m}(motiv_m \cdot gain_m)$$

$$+ \, success\left(\sum_{k} releaser_{goal,k}\right) \cdot (LoI - frustration) + bias \quad (9.1)$$

A_{child} is the activation level of the child behavior group, if present
n is the number of releaser inputs, $releaser_n$
$gain_n$ is the weight for each contributing releaser
m is the number of motivation inputs, $motiv_m$

motiv$_m$ corresponds to the inputs from `drives` or `emotions`

gain$_m$ is the weight for each contributing `drive` or `emotion`

success( ) is a function that returns 1 if the goal has not been achieved, and 0 otherwise

releaser$_{goal.k}$ is a releaser that is active when the goal state is true (i.e., a goal releaser)

LoI is the level of interest, $LoI = LoI_{initial} - decay(LoI, gain_{decayLoI})$

LoI$_{initial}$ is the default persistence

frustration increases linearly with time, $frustration = frustration + (gain_{frust} \cdot t)$

bias is a constant that pre-potentiates the behavior

$decay(x, g) = x - \frac{x}{g}$ for $g > 1$ and $x > 0$, and 0 otherwise

When the behavior group is inactive, the activation level is updated by the equation:

$$A_{behavior} = max \left(A_{child}, \sum_n (releaser_n \cdot gain_n), decay(A_{behavior}, gain_{decayBeh}) \right) \qquad (9.2)$$

Internal Measures

The goal of each behavior is defined as a particular relationship between the robot and its environment (a *goal releaser*). The success condition can simply be represented as another releaser for the behavior that fires when the desired relation is achieved within the appropriate behavioral and motivational context. For instance, the goal condition for the `seek-person` behavior is the `found-person` releaser, which only fires when people are the desired stimulus (the `social-drive` is active), the robot is engaged in a person-finding behavior, and there is a visible person (i.e., skin tone object) who is within face-to-face interaction distance of the robot and is not moving in a threatening manner (no excessive motion). Some behaviors, particularly those at the top level of the hierarchy, operate to maintain a desired internal state (keeping its drive in homeostatic balance, for instance). A releaser for this type of process measures the activation level of the affiliated drive.

The active behavior sends information to the high-level perceptual system that may be needed to provide context for the incoming perceptual features. When a behavior is active, it updates its own internal measures of success and progress to its goal. The behavior sends positive valence to the `emotion` system upon success of the behavior. As time passes with delayed success, an internal measure of `frustration` grows linearly with time. As this grows, it sends negative valence and withdrawn-stance values to the `emotion` system (however, the arousal and stance values may vary as a function of time for some behaviors). The longer it takes the behavior to succeed, the more frustrated the robot appears. The `frustration` level reduces the `level-of-interest` of the behavior. Eventually, the behavior "gives up" and loses the competition to another.

Specificity of Releasers

Behaviors that are located deeper within the hierarchy are more specific. As a result, both the antecedent conditions that release the behavior, as well as the goal relations that signal success, become more specific. This establishes a hierarchy of releasers, progressing in detail from broad and general to more specific. The broadest releasers simply establish the type of stimulus (people versus toys) and its presence or absence. Deeper in the hierarchy, many of the releasers are the same as those that are passed to the affective tagging process in the emotion system. Hence, these releasers are not just simple combinations of perceptual features. They are contextualized according to the motivational and behavioral state of the robot (see chapter 8). They are analogous to simple cognitions in emotional appraisal theories because they specifically relate the perceptual features to the "well-being" and goals of the robot.

Adjustment Parameters

Each behavior follows this general model. Several parameters are used to specify the distinguishing properties of each behavior. This amount of flexibility allows rich behaviors to be specified and interesting behavioral dynamics to be established.

Activation within a group   One important parameter is the releaser used to elicit the behavior. This plays an important role in determining when the behavior becomes active. For instance, the absence of a desired toy stimulus is the correct condition to activate the seek-toy behavior. However, as discussed previously, it is not a simple one-to-one mapping from stimulus to response. Motivational factors also influence a behavior's relevance.

Deactivation within a group   Another important parameter is the goal-signaling releaser. This determines when an appetitive behavior has achieved its goal and can be deactivated. The consummatory behaviors remain active upon success until a motivational switch occurs that biases the robot to tend to a different need. For instance, during the seek-toy behavior (an appetitive behavior), the behavior is successful when the found-toy releaser fires. This releaser is a combination of toy-present with the context provided by the seek-toy behavior. It fires for the short period of time between the decay of the seek-toy behavior and the activation of engage-toy (the consummatory behavior).

Temporal dynamics within a group   The timing of activating and deactivating behaviors within a group is very important. The human and the robot establish a tightly coupled dynamic when in face-to-face interaction. Both are continuously adapting their behavior to the other, and the manner in which they adapt their behavior is often in direct response to the last action the partner just performed. To keep the flow of interaction smooth, the dynamics of behavioral transitions must be well-matched to natural human interaction speeds. For instance, the transition from the `call-to-person` behavior (to bring a distant person near) to the activation of the `greet-person` response (when the person closes to face-to-face interaction distance) to the transition to the `vocal-play` behavior (when the person says his/her first utterance) must occur at a pace that the human feels comfortable with. Each of these involves showing the right amount of responsiveness to the new stimulus situation, the right amount of persistence of the active behavior (the motor act must have enough time to be displayed and witnessed), and the right amount of delay before the next behavior becomes active (so that each display is presented as a purposeful and distinct act).

Temporal dynamics between levels   A similar issue holds for the dynamics between different levels of the hierarchy. If a child behavior is successfully addressing the goal of its parent, then the parent should remain active longer to support the favorable progress of its child. For instance, if the robot is having a good interaction with a person, then the time spent doing so should be extended—rather than rigidly following a fixed schedule where the robot must switch to look for a toy after a certain amount of time. Good quality interactions should not be needlessly interrupted; the timing to address the robot's various needs should be flexible and opportunistic. To accomplish this, the parent behaviors are made aware of the progress of their children. The container node of the child passes activation energy up the hierarchy to its parent, and the parent's activation is a combination of its own measure of relevance and that of its child.

Affective influence   Another important set of parameters adjust how strongly the active behaviors influence the net affective state. The amount of valence, arousal, and stance sent to the `emotion` system can vary from behavior to behavior. Currently, only the leaf behaviors of the hierarchy influence the `emotion` system. Their magnitude and growth rate determine how quickly the robot displays frustration, how strongly it displays pleasure upon success, etc. The timing of affective expression is important, since it often occurs during the transition between different behaviors. Because these affective expressions are social cues, they must occur at the right time to signal the appropriate event that elicited the expression.

For instance, consider the period of time between successfully finding a toy during the `seek-toy` behavior, and the transition to the `engage-toy` behavior. During this time span, the `seek-toy` behavior signals its success to the emotion system by sending it a positively valenced signal. This increase in net positive valence is usually sufficient to cause `joy` to become active, and the robot smiles. The smile is a social cue to the caregiver that the robot has successfully found what it was looking for.

9.4   Kismet's Proto-Social Responses

In the current implementation of the behavior system there are three primary branches, each specialized for addressing a different need. Each is comprised of multiple levels, with three layers being the deepest (see figure 9.2). Each level of the hierarchy serves a different function and addresses a different set of issues. As one moves down in depth, the behaviors serve to more finely tune the relation between the robot and its environment, and in particular, the relation between the robot and the human (Breazeal & Scassellati, 2000).

Level Zero: The Functional Level

The top level of the hierarchy consists of a single behavior group with three behaviors `satiate-social`, `satiate-stimulation`, and `satiate-fatigue` (see figure 9.4). The purpose of this group is to determine which need the robot should address—specifically,

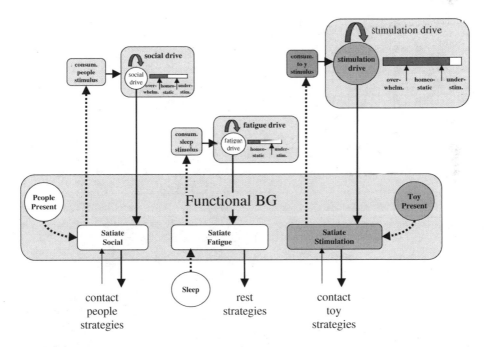

Figure 9.4
The Level Zero behavior group. This is the functional level that establishes which "need" Kismet's behavior will be directed toward satiating. Here, the `stimulation-drive` has the greatest intensity. Furthermore, its satiatory stimulus is present and the `toy-present` releaser is firing. As a result, the `satiate-stimulation` behavior is active and passes the activation from the `toy-present` releaser to satiate the `drive`.

whether the robot should engage people and satiate the `social-drive`, engage toys and satiate the `stimulation-drive`, or rest and satiate the `fatigue-drive`.

To make this decision, each behavior receives input from its affiliated drive. The larger the magnitude of the drive, the more urgently that need must be addressed, and the greater the contribution the `drive` makes to the activation of the behavior. The `satiate-social` behavior receives input from the `people-present` releaser, and the `satiate-stimulation` behavior receives input from the `toy-present` releaser. The value of each of these releasers is proportional to the intensity of the associated stimulus (for instance, closer objects appear larger in the visual field and have a higher releaser value). The `fatigue-drive` is somewhat different; it receives input from the activation of the `sleep` behavior.

The winning behavior at this level performs two functions. First, it spreads activation downward to the next level of the hierarchy. Thus, behavior becomes organized around satisfying the affiliated `drive`. This establishes the motivational context that determines whether a given type of stimulus is desirable (whether it satiates the affiliated `drive` of the active behavior).

Second, the top-level behaviors act to satiate their affiliated `drives`. Each satiates its `drive` when the robot encounters a good-intensity stimulus (neither under-stimulating nor overwhelming). "Satiation" moves the `drive` to the homeostatic regime. If the stimulus is too intense, the `drive` moves to the overwhelmed regime. If the stimulus is not intense enough, the drive moves to the under-stimulated regime. These conditions are addressed by Level One behaviors.

Level One: The Environment-Regulation Level

The behaviors at this level are responsible for establishing a good intensity of interaction with the environment (see figure 9.5). The behaviors `satiate-social` and `satiate-stimulation` each pass activation to their Level One behavior group below. The behavior group consists of three types of behaviors: *searching* behaviors set the current task to explore the environment and to bring the robot into contact with the desired stimulus; *avoidance* behaviors set the task to move the robot away from stimuli that are too intense, undesirable, or threatening; and *engagement* behaviors set the task of interacting with desirable, good-intensity stimuli.

Search behavior establishes the goal of finding the desired stimuli. Thus, the goal of the `seek-people` behavior is to seek out skin-toned stimuli, and the goal of the `seek-toys` behavior is to seek out colorful stimuli. As described in chapter 6, an active behavior adjusts the gains of the attention system to facilitate these goals. Each search behavior receives contributions from releasers (signaling the absence of the desired stimulus) or low arousal affective states (such as `boredom` and `sorrow`) that signal a prolonged absence of the sought-after stimulus.

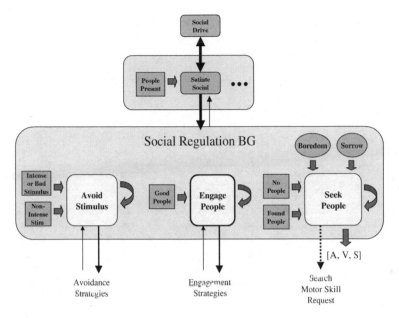

Figure 9.5
Level One behavior group. Only the social hierarchy is shown. This is the environment-regulation level that establishes interactions that neither under-stimulate nor overwhelm the robot.

Avoidance behavior, avoid-stimulus for both the social and stimulation hierarchies, establishes the goal of putting distance between the robot and the offending stimulus or event. The presence of an offensive stimulus or event contributes to the activation of an avoidance behavior through its releaser. At this level, an offending stimulus is either "undesirable" (not of the correct type), "threatening" (very close and moving fast), or "annoying" (too close or moving too fast to be visually tracked effectively). The behavioral response recruited to cope with the situation depends on the nature of the offense. The coping strategy is defined within the behavior group one more level down. The specifics of Level Two are discussed below.

The goal of the engagement behaviors, engage-people or engage-toys, is to orient and maintain the robot's attention on the desired stimulus. These are the consummatory behaviors of the Level One group. With the desired stimulus found, and any offensive conditions removed, the robot can engage in play behaviors with the desired stimulus. These play behaviors are described later in this section.

Level Two: The Protective Behaviors

As shown in figure 9.6, there are three types of protective behaviors that co-exist within the Protective Level Two behavior group. Each represents a different coping strategy

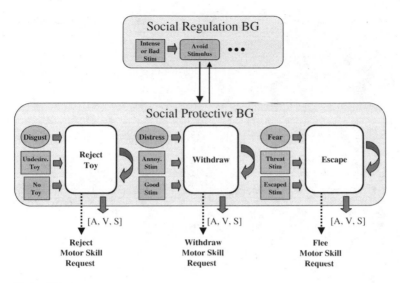

Figure 9.6
Level Two protective behavior group. Only the social hierarchy is shown. This is the level two behavior group that allows the robot to avoid offensive stimuli. See text.

that is responsible for handling a particular kind of offense. Each coping strategy receives contributions from its affiliated releaser as well as from its affiliated emotion process.

When active, the goal set by the `escape` behavior is to flee from the offending stimulus. This behavior sends a request to the motor system to perform the fleeing response, where the robot closes its eyes, grimaces, and turns its head away from a threatening stimulus. It doesn't matter whether this stimulus is skin-toned or colorful—if anything is very close and moving fast, it is interpreted as a threat by the low-level visual perception system. There is a dedicated releaser, `threat-stimulus`, that fires whenever a threatening stimulus is encountered. This releaser passes activation to the `escape` behavior as well as to the `emotion` system. When `fear` is active, it elicits a fearful expression on the robot's face of the appropriate intensity (see chapters 8 and 10). This expression is a social signal that gives advance warning of any behavioral response that may ensue. If the activation level of `fear` is strong enough, it sends sufficient activation to the `escape` behavior to win the competition. The robot then performs the escape maneuver. A few of these behaviors can be viewed in the "Emotive Responses" section of the included CD-ROM.

The `withdraw` behavior is active when the robot finds itself in an unpleasant, but not threatening, situation. Often this corresponds to a situation where the robot's visual processing abilities are over-challenged. For instance, if a person is too close to the robot, the eye-detector has difficulty locating the person's eyes. Alternatively, if a person is waving a toy too fast to be tracked effectively, the excessive amount of motion is classified as "annoy-ing" by the low-level visual processes. Either of these conditions will cause the `annoy-stim` releaser to fire. The releaser sends activation energy to the `withdraw` behavior as well as to the `emotion` system. This causes the `distress` process to become active. Once active, the robot's face exhibits an annoyed appearance. `Distress` also sends sufficient activation to activate the `withdraw` behavior, and a request is made of the motor system to back away from the offending stimulus. The primary function of this response is to send a social cue to the human that they are offending the robot and thereby encourage the person to modify her behavior.

The `reject` behavior is active when the robot is being offered an undesirable stimulus. The affiliated emotion process is `disgust`. It is similar to the situation where an infant will not accept the food it is offered. It has nothing to do with the offered stimulus being noxious, it is simply not what the robot is after.

Level Two: The Play Behaviors

Kismet exhibits different play patterns when engaging toys versus people. Kismet will readily track and occasionally vocalize while its attention is drawn to a colorful toy, but it will not evoke its repertoire of envelope displays that characterize vocal play. These proto-dialogue behaviors are reserved for interactions with people. These social cues are not exhibited when playing with toys. The difference in the manner Kismet interacts with people versus toys provides observable evidence that these two categories of stimuli are distinguished by Kismet.

In this section I focus the discussion on those four behaviors within the `Social Play Level Two` behavior group. This behavior group encapsulates Kismet's engagement strategies for establishing proto-dialogues during face-to-face exchanges. They finely tune the relation between the robot and the human to support interactive games at a level where both partners perform well.

The first engagement task is the `call-to-person` behavior. This behavior is relevant when a person is in view of the robot but too far away for face-to-face exchange. The goal of the behavior is to lure the person into face-to-face interaction range (ideally, about three feet from the robot). To accomplish this, Kismet sends a social cue, the calling display, directed to the person within calling range. A demonstration of this behavior is viewable on the CD-ROM in the section titled "Social Amplification."

The releaser affiliated with this behavior combines skin-tone with proximity measures. It fires when the person is four to seven feet from the robot. The actual calling display is covered in detail in chapter 10. It is evoked when the `call-to-person` behavior is active and makes a request to the motor system to exhibit the display. The human observer sees the robot orient toward him/her, crane its neck forward, wiggle its ears with large amplitude movements, and vocalize excitedly. The display is designed to attract a person's attention. The robot then resumes a neutral posture, perks its ears, and raises its brows in an expectant manner. It waits in this posture for a while, giving the person time to approach before the calling sequence resumes. The `call-to-person` behavior will continue to request the display from the motor system until it is either successful and becomes deactivated, or it becomes irrelevant.

The second task is the `greet-person` behavior. This behavior is relevant when the person has just entered face-to-face interaction range. It is also relevant if the Social Play Level Two behavior group has just become active and a person is already within face-to-face range. The goal of the behavior is to socially acknowledge the human and to initiate a close interaction. When active, it makes a request of the motor system to perform the greeting display. The display involves making eye contact with the person and smiling at them while waving the ears gently. It often immediately follows the success of the `call-to-person` behavior. It is a transient response, only issued once, as its completion signals the success of this behavior.

The third task is `attentive-regard`. This behavior is active when the person has already established a good face-to-face interaction distance with the robot but remains silent. The goal of the behavior is to visually attend to the person and to appear open to interaction. To accomplish this, it sends a request to the motor system to hold gaze on the person, ideally looking into the person's eyes if the eye detector can locate them. The robot watches the person intently and vocalizes occasionally. If the person does speak, this behavior loses the competition to the `vocal-play` behavior. This behavior is viewable on the CD-ROM in the fifth demonstration, "Visual Behaviors."

The fourth task is `vocal-play`. The goal of this behavior is to carry out a proto-dialogue with the person. It is relevant when the person is within face-to-face interaction distance and has spoken. To perform this task successfully, the `vocal-play` behavior must closely regulate turn-taking with the human. This involves a close interaction with the perceptual system to perceive the relevant turn-taking cues from the person (i.e., that a person is present and whether there is speech occurring), and with the motor system to send the relevant turn-taking cues back to the person. Video demonstrations of Kismet's "Proto-Conversations" can be viewed on the accompanying CD-ROM.

There are four turn-taking phases this behavior must recognize and respond to. Each state is recognized using distinct perceptual cues, and each phase involves making specific display requests of the motor system:

• *Relinquish speaking turn* This phase is entered immediately after the robot finishes speaking. The robot relinquishes its turn by craning its neck forward, raising its brows, and making eye-contact (in adult humans, shifting gaze direction is sufficient, but Kismet's display is exaggerated to increase readability). It holds its gaze on the person throughout this phase. Due to noise in the visual system, however, the eyes tend to flit about the person's face, perhaps even leaving it briefly and then returning soon afterwards. This display signals that the robot has finished speaking and is waiting for the human to say something. It will time out after approximately 8 seconds if the person does not respond. At this point, the robot reacquires its turn and issues another vocalization in an attempt to reinitiate the dialogue.

• *Attend to human's speech* Once the perceptual system acknowledges that the human has started speaking, the robot's ears perk. This subtle feedback cue signals that the robot is listening to the person speak. The robot looks generally attentive to the person and continues to maintain eye contact if possible.

• *Reacquire speaking turn* This phase is entered when the perceptual system acknowledges that the person's speech has ended. The robot signals that it is about to speak by leaning back to a neutral posture and averting its gaze. The robot is likely to blink its eyes as it shifts posture.

• *Deliver speech* Soon after the robot shifts its posture back to neutral, the robot vocalizes. The utterances are short babbles, generated by the vocalization system (presented in chapter 11). Sometimes more than one is issued. The eyes migrate back to the person's face, to their eyes if possible. Just before the robot is prepared to finish this phase, it is likely to blink. The behavior transitions back to the relinquish turn phase and the cycle resumes.

The system is designed to maintain social exchanges with a person for about twenty minutes; at this point the other `drives` typically begin to dominate the robot's motivation. When this occurs, the robot begins to behave in a fussy manner—the robot becomes more distracted by other things around it, and it makes fussy faces more frequently. It is more difficult to engage in proto-dialogue. Overall, it is a significant change in behavior. People seem to sense the change readily and try to vary the interaction, often by introducing a toy. The smile that appears on the robot's face and the level of attention that it pays to the toy are strong cues that the robot is now involved in satiating its `stimulation-drive`.

9.5 Overview of the Motor Systems

Whereas the behavior system is responsible for deciding which task the robot should perform at any time, the motor system is responsible for figuring out how to drive the motors in order to carry out the task. In addition, whereas the motivation system is responsible for establishing the affective state of the robot, the motor system is responsible for commanding the actuators in order to convey that emotional state.

There are four distinct motor systems that carry out these functions for Kismet. The vocalization system produces expressive babbles that allow the robot to engage humans in proto-dialogue. The face motor system orchestrates the robot's emotive facial expressions and body posture, its facial displays that serve communicative social functions, those that serve behavioral functions (such as "sleeping"), and lip synchronization with accompanying facial animation. The oculo-motor system produces human-like eye movements and head orientations that serve important sensing as well as social functions. Finally, the motor skills system coordinates each of these specialized motor systems to produce coherent multi-modal motor acts.

Levels of Interaction

Kismet's rich motor behavior can be conceptualized on four different levels (as shown in figure 9.7). These levels correspond to the *social level,* the *behavior level,* the *skills level,* and the *primitives level.* This decomposition is motivated by distinct temporal, perceptual, and interaction constraints at each level.

The temporal constraints pertain to how fast the motor acts must be updated and executed. These can range from real-time vision rates (33 frames/sec) to the relatively slow time-scale of social interaction (potentially transitioning over minutes).

The perceptual constraints pertain to what level of sensory feedback is required to co-ordinate behavior at that layer. This perceptual feedback can originate from the low-level

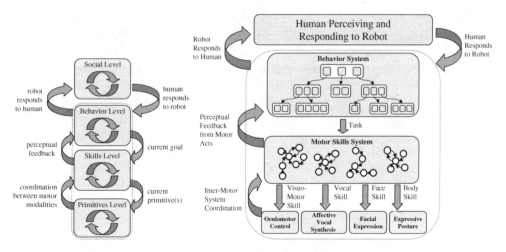

Figure 9.7
Levels of behavioral organization. The primitive level is populated with tightly coupled sensori-motor loops. The skill level contains modules that coordinate primitives to achieve tasks. Behavior level modules deal with questions of relevance, persistence and opportunism in the arbitration of tasks. The social level comprises design-time considerations of how the robot's behaviors will be interpreted and responded to in a social environment.

visual processes, such as the current target from the attention system, to relatively high-level multi-modal percepts generated by the behavioral releasers.

The interaction constraints pertain to the arbitration of units that compose each layer. This can range from low-level oculo-motor primitives (such as saccades and smooth pursuit) to using visual behavior to regulate turn-taking.

Each level serves a particular purpose for generating the overall observed behavior. As such, each level must address a specific set of issues. The levels of abstraction help simplify the overall control of behavior by restricting each level to address those core issues that are best managed at that level. By doing so, the coordination of behavior at each level (i.e., arbitration), between the levels (i.e., top-down and bottom-up), and through the world is maintained in a principled way.

The social level explicitly deals with issues pertaining to having a human in the interaction loop. This requires careful consideration of how the human interprets and responds to the robot's behavior in a social context. Using visual behavior (making eye contact and breaking eye contact) to help regulate the transition of speaker turns during vocal turn-taking is an example presented in chapter 9. Chapter 7 discusses examples with respect to affect-based interactions during "emotive" vocal exchanges. Chapter 12 discusses the relationship between animate visual behavior and social interaction. A summary of these findings is presented in chapter 13.

The behavior level deals with issues related to producing relevant, appropriately persistent, and opportunistic behavior. This involves arbitrating between the many possible goal-achieving behaviors that Kismet could perform to establish the current task. Actively seeking out a desired stimulus and then visually engaging it is an example. Other behavior examples are described in chapter 9.

The motor skills level is responsible for figuring out how to move the motors to accomplish the task specified by the behavior system. Fundamentally, this level deals with the blending of and sequencing between coordinated ensembles of motor primitives (each ensemble is a distinct motor skill). The skills level must also deal with coordinating multi-modal motor skills (e.g., those motor skills that combine speech, facial expression, and body posture). Kismet's searching behavior is an example where the robot alternately performs ballistic eye-neck orientation movements with gaze fixation to the most salient target. The ballistic movements are important for scanning the scene, and the fixation periods are important for locking on the desired type of stimulus. I elaborate upon this system at the end of this chapter.

The motor primitives level implements the building blocks of motor action. This level must deal with motor resource allocation and tightly coupled sensori-motor loops. Kismet actually has three distinct motor systems at the primitives level: the *expressive vocal system* (see chapter 11), the *facial animation system* (see chapter 10), the *oculo-motor system* (see chapter 12). Aspects of controlling the robot's body posture are described in chapters 10 and 12.

The Motor Skills System

Given the current task (as dictated by the behavior system), the motor skills system is responsible for figuring out how to carry out the stated goal. Often this requires coordinating multiple motor modalities (speech, body posture, facial display, and gaze control). Requests for these modalities can originate from the top down (i.e., from the `emotion` system or behavior system) as well as from the bottom-up (e.g., the vocal system requesting lip and jaw movements for lip synchronizing). Hence, the motor skills level must address the issue of servicing the motor requests of different systems across the different motor resources.

The motor skills system also must appropriately blend the motor actions of concurrently active behaviors. Sometimes concurrent behaviors require completely different sets of actuators (such as babbling while watching a stimulus). In this case there is no direct competition over a shared resource, so the motor skills system should command the actuators to execute both behaviors simultaneously. Other times, two concurrently active behaviors may compete for the same actuators. For instance, the robot may have to smoothly track a moving object while maintaining vergence. These two behaviors are complementary in that each can be carried out without the sacrifice or degradation in the performance of the other. However, the motor skills system must coordinate the motor commands to do so appropriately.

The motor skills system is also responsible for smoothly transitioning between sequentially active behaviors. For instance, to initiate a social exchange, the robot must first mutually orient to the caregiver and then exchange a greeting with her. Once started, Kismet may take turns with the caregiver in exchanging vocalizations, facial expressions, etc. After a while, either party can disengage from the other (such as by looking away), thereby terminating the interaction. While sequencing between these behaviors, the motor system must figure out how to transition smoothly between them in a timely manner so as not to disrupt the natural flow of the interaction.

Finally, the motor skills system is responsible for moving the robot's actuators to convey the appropriate emotional state of the robot. This may involve performing facial expressions, or adapting the robot's posture. Of course, this affective state must be conveyed while carrying out the active task(s). This is a special case of blending mentioned above, which may or may not compete for the same actuators. For instance, looking at an unpleasant stimulus may be performed by directing the eyes to the stimulus, but orienting the face away from the stimulus and configuring the face into a "disgusted" look.

Motor Skill Mechanisms

It often requires a sequence of coordinated motor movements to satisfy a goal. Each motor movement is a primitive (or a combination of primitives) from one of the base motor systems (the vocal system, the oculo-motor system, etc.). Each of these coordinated series of motor primitives is called a *skill,* and each skill is implemented as a finite state machine (FSM). Each motor skill encodes knowledge of how to move from one motor state to the next, where each sequence is designed to bring the robot closer to the current goal. The motor skills level must arbitrate among the many different FSMs, selecting the one to become active based on the active goal. This decision process is straightforward since there is an FSM tailored for each task of the behavior system.

Many skills can be thought of as *fixed action patterns* (FAPs) as conceptualized by early ethologists (Tinbergen, 1951; Lorenz, 1973). Each FAP consists of two components, the *action* component and the *taxis* (or orienting) component. For Kismet, FAPs often correspond to communicative gestures where the action component corresponds to the facial gesture, and the taxis component (to whom the gesture is directed) is controlled by gaze. People seem to intuitively understand that when Kismet makes eye contact with them, they are the locus of Kismet's attention and the robot's behavior is organized about them. This places the person in a state of action readiness where they are poised to respond to Kismet's gestures.

A classic example of a motor skill is Kismet's calling FAP (see figure 9.8). When the current task is to bring a person into a good interaction distance, the motor skill system activates the `calling` FSM. The taxis component of the FAP issues a `hold gaze` request to the oculo-motor system. This serves to maintain the robot's gaze on the person. In the first state (1) of the gesture component, Kismet leans its body toward the person (a request to the body posture motor system). This strengthens the person's perception that the robot has taken a particular interest in them. The ears also begin to waggle exuberantly (creating a significant amount of motion and noise) which further attracts the person's attention to

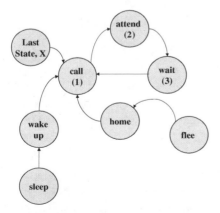

Figure 9.8
The calling motor skill. The states 1, 2, and 3 are described in the text. The remaining states encode knowledge of how to transition from any previously active motor skill state to the `call` state.

the robot. In addition, Kismet vocalizes excitedly, which is perceived as an initiation. The FSM transitions to the second state (2) upon the completion of this gesture. In this state, the robot "sits back" and waits for a bit with an expectant expression (ears slightly perked, eyes slightly widened, and brows raised). If the person has not already approached the robot, it is likely to occur during this "anticipation" phase. If the person does not approach within the allotted time period, the FSM transitions to the third state (3) where face relaxes, the robot maintains a neutral posture, and gaze fixation is released. At this point, the robot is able to shift gaze. As long as this FSM is active (determined by the behavior system), the calling cycle repeats. It can be interrupted at any state transition by the activation of another FSM (such as the `greeting` FSM when the person has approached). Chapter 10 presents a table and summary of FAPs that have been implemented on Kismet.

9.6   Playful Interactions with Kismet

The behavior system implements the four classes of proto-social responses. The robot displays affective responses by changing emotive facial expressions in response to stimulus quality and internal state. These expressions relate to goal achievement, emotive reactions, and reflections of the robot's state of "well-being." The exploratory responses include visual search for desired stimuli, orientation, and maintenance of mutual regard. Kismet has a variety of protective responses that serve to distance the robot from offending stimuli. Finally, the robot has a variety of regulatory responses that bias the caregiver to provide the appropriate level and kinds of interactions at the appropriate times. These are communicated to the caregiver through carefully timed social displays as well as affective facial expressions. The organization of the behavior system addresses the issues of relevancy, coherency, persistence, flexibility, and opportunism. The proto-social responses address the issues of believability, promoting empathy, expressiveness, and conveying intentionality.

Regulating Interaction

Figure 9.9 shows Kismet responding to a toy with these four response types. The robot begins the trial looking for a toy and displaying sadness (an affective response). The robot immediately begins to move its eyes searching for a colorful toy stimulus (an exploratory response) ($t < 10$). When the caregiver presents a toy ($t \approx 13$), the robot engages in a play behavior and the `stimulation-drive` becomes satiated ($t \approx 20$). As the caregiver moves the toy back and forth ($20 < t < 35$), the robot moves its eyes and neck to maintain the toy within its field of view. When the stimulation becomes excessive ($t \approx 35$), the robot becomes first "displeased" and then "fearful" as the `stimulation-drive` moves into the overwhelmed regime. After extreme over-stimulation, a protective escape response produces a large neck movement ($t = 38$), which removes the toy from the field of view.

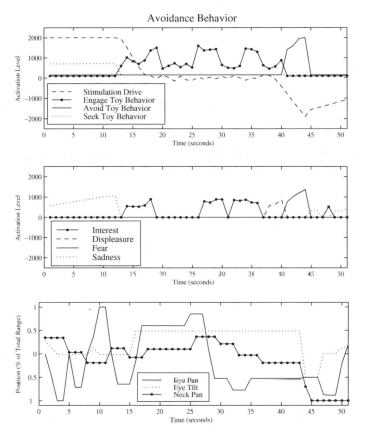

Figure 9.9
Kismet's response to excessive stimulation. Behaviors and `drives` (top), `emotions` (middle), and motor output (bottom) are plotted for a single trial of approximately 50 seconds.

Once the stimulus has been removed, the `stimulation-drive` begins to drift back to the homeostatic regime (one of the many regulatory responses in this example).

Interaction Dynamics

The behavior system produces interaction dynamics that are similar to the five phases of infant social interactions (initiation, mutual-orientation, greeting, play-dialogue, and disengagement) discussed in chapter 3. These dynamic phases are not explicitly represented in the behavior system, but emerge from the interaction of the synthetic nervous system with the environment. Producing behaviors that convey intentionality exploits the caregiver's natural tendencies to treat the robot as a social creature, and thus to respond in characteristic

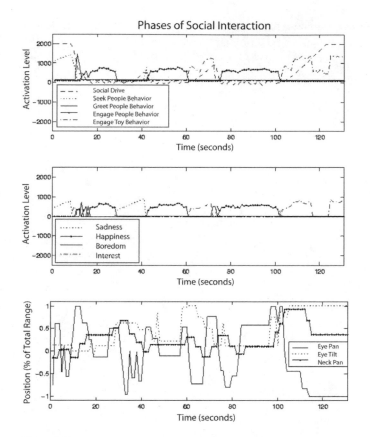

Figure 9.10
Cyclic responses during social interaction. Behaviors and `drives` (top), `emotions` (middle), and motor output (bottom) are plotted for a single trial of approximately 130 seconds.

ways to the robot's overtures. This reliance on the external world produces dynamic behavior that is both flexible and robust.

Figure 9.10 shows Kismet's dynamic responses during face-to-face interaction with a caregiver. Kismet is initially looking for a person and displaying sadness (the initiation phase). The sad expression evokes nurturing responses from the caregiver. The robot begins moving its eyes looking for a face stimulus ($t < 8$). When it finds the caregiver's face, it makes a large eye movement to enter into mutual regard ($t \approx 10$). Once the face is foveated, the robot displays a greeting behavior by wiggling its ears ($t \approx 11$) and begins a play-dialogue phase of interaction with the caregiver ($t > 12$). Kismet continues to engage the caregiver until the caregiver moves outside the field of view ($t \approx 28$). Kismet quickly becomes "sad" and begins to search for a face, which it re-acquires when the caregiver returns ($t \approx 42$). Eventually, the robot habituates to the interaction with the caregiver and begins to attend to a toy that the caregiver has provided ($60 < t < 75$). While interacting with the toy, the robot displays interest and moves its eyes to follow the moving toy. Kismet soon habituates to this stimulus and returns to its play-dialogue with the caregiver ($75 < t < 100$). A final disengagement phase occurs ($t \approx 100$) when the robot's attention shifts back to the toy.

Regulating Vocal Exchanges

Kismet employs different social cues to regulate the rate of vocal exchanges. These include both eye movements as well as postural and facial displays. These cues encourage the subjects to slow down and shorten their speech. This benefits the auditory processing capabilities of the robot.

To investigate Kismet's performance in engaging people in proto-dialogues, I invited three naive subjects to interact with Kismet. They ranged in age from 25 to 28 years of age. There were one male and two females, all professionals. They were asked simply to talk to the robot. Their interactions were videorecorded for further analysis. (Similar video interactions can be viewed on the accompanying CD-ROM.)

Often the subjects begin the session by speaking longer phrases and only using the robot's vocal behavior to gauge their speaking turn. They also expect the robot to respond immediately after they finish talking. Within the first couple of exchanges, they may notice that the robot interrupts them, and they begin to adapt to Kismet's rate. They start to use shorter phrases, wait longer for the robot to respond, and more carefully watch the robot's turn-taking cues. The robot prompts the other for his/her turn by craning its neck forward, raising its brows, and looking at the person's face when it's ready for him/her to speak. It will hold this posture for a few seconds until the person responds. Often, within a second of this display, the subject does so. The robot then leans back to a neutral posture, assumes a neutral expression, and tends to shift its gaze away from the person. This cue indicates that the robot is about to speak. The robot typically issues one utterance, but it may issue several. Nonetheless, as the exchange proceeds, the subjects tend to wait until prompted.

Before the subjects adapt their behavior to the robot's capabilities, the robot is more likely to interrupt them. There tends to be more frequent delays in the flow of "conversation," where the human prompts the robot again for a response. Often these "hiccups" in the flow appear in short clusters of mutual interruptions and pauses (often over two to four speaking turns) before the turns become coordinated and the flow smoothes out. By analyzing the video of these human-robot "conversations," there is evidence that people entrain to the robot (see table 9.1). These "hiccups" become less frequent. The human and robot are able to carry on longer sequences of clean turn transitions. At this point the rate of vocal exchange is well-matched to the robot's perceptual limitations. The vocal exchange is reasonably fluid.

Table 9.1
Data illustrating evidence for entrainment of human to robot.

		Time Stamp (min:sec)	Time Between Disturbances (sec)
subject 1	start 15:20	15:20–15:33	13
		15:37–15:54	21
		15:56–16:15	19
		16:20–17:25	70
	end 18:07	17:30–18:07	37+
subject 2	start 6:43	6:43–6:50	7
		6:54–7:15	21
		7:18–8:02	44
	end 8:43	8:06–8:43	37+
subject 3	start 4:52	4:52–4:58	10
		5:08–5:23	15
		5:30–5:54	24
		6:00–6:53	53
		6:58–7:16	18
		7:18–8:16	58
		8:25–9:10	45
	end 10:40	9:20–10:40	80+

Table 9.2
Kismet's turn-taking performance during proto-dialogue with three naive subjects. Significant disturbances are small clusters of pauses and interruptions between Kismet and the subject until turn-taking becomes coordinated again.

	Subject 1		Subject 2		Subject 3		
	Data	Percent	Data	Percent	Data	Percent	Average
Clean Turns	35	83	45	85	83	78	82
Interrupts	4	10	4	7.5	16	15	11
Prompts	3	7	4	7.5	7	7	7
Significant Flow Disturbances	3	7	3	5.7	7	7	6.5
Total Speaking Turns	42		53		106		

Table 9.2 shows that the robot is engaged in a smooth proto-dialogue with the human partner the majority of the time (about 82 percent).

9.7   Limitations and Extensions

Kismet can engage a human in compelling social interaction, both with toys and during face-to-face exchange. People seem to interpret Kismet's emotive responses quite naturally and adjust their behavior so that it is suitable for the robot. Furthermore, people seem to entrain to the robot by reading its turn-taking cues. The resulting interaction dynamics are reminiscent of infant-caregiver exchanges. However, there are number of ways in which the system could be improved.

The robot does not currently have the ability to interrupt itself. This will be an important ability for more sophisticated exchanges. When watching video of people talking with Kismet, they are quite resilient to hiccups in the flow of "conversation." If they begin to say something just before the robot, they will immediately pause once the robot starts speaking and wait for the robot to finish. It would be nice if Kismet could exhibit the same courtesy. The robot's babbles are quite short at the moment, so this is not a serious issue yet. As the utterances become longer, it will become more important.

It is also important for the robot to understand where the human's attention is directed. At the very least, the robot should have a robust way of measuring when a person is addressing it. Currently the robot assumes that if a person is nearby, then that person is attending to the robot. The robot also assumes that it is the most salient person who is addressing it. Clearly this is not always the case. This is painfully evident when two people try to talk to the robot and to each other. It would be a tremendous improvement to the current implementation if the robot would only respond when a person addressed it directly (instead of addressing someone else) and if the robot responded to the correct person (instead of the most salient person). Sound localization using the stereo microphones on the ears could help identify the source of the speech signal. This information could also be correlated with visual input to direct the robot's gaze. In general, determining where a person is looking is a computationally difficult problem (Newman & Zelinsky, 1998; Scassellati, 1999).

The latency in Kismet's verbal turn-taking behavior needs to be reduced. For humans, the average time for a verbal reply is about 250 ms. For Kismet, its verbal response time varies from 500 ms to 1500 ms. Much of this depends on the length of the person's previous utterance, and the time it takes the robot to shift between turn-taking postures. In the current implementation, the in-speech flag is set when the person begins speaking, and is cleared when the person finishes. There is a delay of about 500 ms built into the speech recognition system from the end of speech to accommodate pauses between phrases. Additional delays are related to the length of the spoken utterance—the longer the utterance the more computation is required before the output is produced. To alleviate awkward pauses and to give people immediate feedback that the robot heard them, the ear-perk response is triggered by the sound-flag. This flag is sent immediately whenever the speech recognizer receives input (speech or non-speech sounds). Delays are also introduced as the robot shifts posture between taking its turn and relinquishing the floor. This also sends important social cues and enlivens the exchange. In watching the video, the turn-taking pace is certainly slower than for conversing adults, but given the lively posturing and facial animation, it appears engaging. The naive subjects readily adapted to this pace and did not seem to find it awkward.

To scale the performance to adult human performance, however, the goal of a 250 ms delay between speaking turns should be achieved.

9.8   Summary

Drawing strong inspiration from ethology, the behavior system arbitrates among competing behaviors to address issues of relevance, coherency, flexibility, robustness, persistence, and opportunism. This enables Kismet to behave in a complex, dynamic world. To socially engage a human, however, its behavior must address issues of believability—such as conveying intentionality, promoting empathy, being expressive, and displaying enough variability to appear unscripted while remaining consistent. To accomplish this, a wide assortment of proto-social, infant-like responses have been implemented. These responses encourage the human caregiver to treat the robot as a young, socially aware creature. Particular attention has been paid to those behaviors that allow the robot to actively engage a human, to call to people if they are too far away, and to carry out proto-dialogues with them when they are nearby. The robot employs turn-taking cues that humans use to entrain to the robot. As a result, the proto-dialogues become smoother over time. The general dynamics of the exchange share structural similarity with those of three-month-old infants with their caregivers. All five phases (initiation, mutual regard, greeting, play dialogue, and disengagement) can be observed.

Kismet's motor behavior is conceptualized, modeled, and implemented on multiple levels. Each level is a layer of abstraction with distinct timing, sensing, and interaction characteristics. Each layer is implemented with a distinct set of mechanisms that address these factors. The motor skills system coordinates the primitives of each specialized system for facial animation, body posture, expressive vocalization, and oculo-motor control. I describe each of these specialized motor systems in detail in the following chapters.

10 Facial Animation and Expression

The human face is the most complex and versatile of all species (Darwin, 1872). For humans, the face is a rich and versatile instrument serving many different functions. It serves as a window to display one's own motivational state. This makes one's behavior more predictable and understandable to others and improves communication (Ekman et al., 1982). The face can be used to supplement verbal communication. A quick facial display can reveal the speaker's attitude about the information being conveyed. Alternatively, the face can be used to complement verbal communication, such as lifting of the eyebrows to lend additional emphasis to a stressed word (Cassell, 1999b). Facial gestures can communicate information on their own, such as a facial shrug to express "I don't know" to another's query. The face can serve a regulatory function to modulate the pace of verbal exchange by providing turn-taking cues (Cassell & Thorisson, 1999). The face serves biological functions as well—closing one's eyes to protect them from a threatening stimulus and, on a longer time scale, to sleep (Redican, 1982).

10.1 Design Issues for Facial Animation

Kismet doesn't engage in adult-level discourse, but its face serves many of these functions at a simpler, pre-linguistic level. Consequently, the robot's facial behavior is fairly complex. It must balance these many functions in a timely, coherent, and appropriate manner. Below, I outline a set of design issues for the control of Kismet's face.

Real-time response   Kismet's face must respond at interactive rates. It must respond in a timely manner to the person who engages it as well to other events in the environment. This promotes readability of the robot, so the person can reliably connect the facial reaction to the event that elicited it. Real-time response is particularly important for sending expressive cues to regulate social dynamics. Excessive latencies disrupt the flow of the interaction.

Coherence   Kismet has fifteen facial actuators, many of which are required for any single emotive expression, behavioral display, or communicative gesture. There must be coherence in how these motor ensembles move together, and how they sequence between other motor ensembles. Sometimes Kismet's facial behaviors require moving multiple degrees of freedom to a fixed posture, sometimes the facial behavior is an animated gesture, and sometimes it is a combination of both. If the face loses coherence, the information it contains is lost to the human observer.

Synchrony   The face is one expressive modality that must work in concert with vocal expression and body posture. Requests for these motor modalities can arise from multiple sources in the synthetic nervous system. Hence, synchrony is an important issue. This is of particular importance for lip synchronization where the phonemes spoken during a vocal utterance must be matched by the corresponding lip postures.

Expressive versatility   Kismet's face currently supports four different functions. It reflects the state of the robot's emotion system, called emotive expressions. It conveys social cues during social interactions with people, called expressive facial displays. It synchronizes with the robot's speech, and it participates in behavioral responses. The face system must be quite versatile as the manner in which these four functions are manifest changes dynamically with motivational state and environmental factors.

Readability   Kismet's face must convey information in a manner as similar to humans as possible. If done sufficiently well, then naive subjects should be able to read Kismet's facial expressions and displays without requiring special training. This fosters natural and intuitive interaction between Kismet and the people who interact with it.

Believability   As with much of Kismet's design, there is a delicate balance between complexity and simplicity. Enforcing levels of abstraction in the control hierarchy with clean interfaces is important for promoting scalability and real-time response. The design of Kismet's face also strives to maintain a balance. It is quite obviously a caricature of a human face (minus the ears!) and therefore cannot do many of the things that human faces do. However, by taking this approach, people's expectations for realism must be lowered to a level that is achievable without detracting from the quality of interaction. As argued in chapter 5, a realistic face would set very high expectations for human-level behavior. Trying to achieve this level of realism is a tremendous engineering challenge currently being attempted by others (Hara, 1998). It is not necessary for the purposes here, however, which focus on natural social interaction.

10.2   Levels of Face Control

The face motor system consists of six subsystems organized into four layers of control. As presented in chapter 9, the face motor system communicates with the motor skill system to coordinate over different motor modalities (voice, body, and eyes). An overview of the face control hierarchy is shown in figure 10.1. Each layer represents a level of abstraction with its own interfaces for communicating with the other levels. The highest layers control ensembles of facial features and are organized by facial function (emotive expression, lip synchronization, facial display). The lowest layer controls the individual degrees of freedom. Enforcing these levels of abstraction keeps the system modular, scalable, and responsive.

The Motor Demon Layer

The lowest level is called the *motor demon* layer. It is organized by individual actuators and implements the interface to access the underlying hardware. It initializes the maximum, minimum, and reference positions of each actuator and places safety caps on them. A

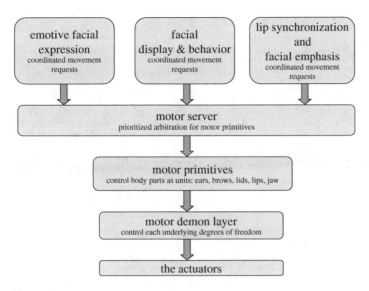

Figure 10.1
Levels of abstraction for facial control.

common reference frame is established for all the degrees of freedom so that values of the same sign command all actuators in a consistent direction. The interface allows other processes to set the position and velocity targets of each actuator. These values are updated in a tight loop 30 times per second. Once these values are updated, the target requests are converted into a pulse-width-modulated control signal. Each is then sent through the *TPU* lines of the 68332 to drive the 14 futaba servo motors. In the case of the jaw, these values are scaled and passed on to QNX where the MEI motion controller card servos the jaw.

The Motor Primitives Layer

The next level up is the *motor primitives* layer. Here, the interface groups the underlying actuators by facial feature. Each motor primitive controls a separate body part (such as an ear, a brow, an eyelid, the upper lip, the lower lip, or the jaw). Higher-level processes make position and velocity requests of each facial feature in terms of their observed movement (as opposed to their underlying mechanical implementation). For instance, the left ear motor primitive converts requests to control elevation, rotation, and speed to the underlying differentially geared motor ensemble. The interface supports both postural movements (go to a specified position) as well as rhythmic movements (oscillate for a number of repetitions with a given speed, amplitude, and period). The interface implements a second set of primitives for small groups of facial features that often move together (such as wiggling both ears, or knitting both brows, or blinking both lids.) These are simply constructed from those primitives controlling each individual facial feature.

The Motor Server Layer

The *motor server layer* arbitrates the requests for facial expression, facial display, or lip synchronization. Requests originating from these three functions involve moving ensembles of facial features in a coordinated manner. These requests are often made concurrently. Hence, this layer is responsible for blending and or sequencing these incoming requests so that the observed behavior is coherent and synchronized with the other motor modalities (voice, eyes, and head).

In some cases, there is blending across orthogonal sets of facial features when subsystems serving different facial functions control different groups of facial features. For instance, when issuing a verbal greeting the lip synchronization process controls the lips and jaw while a facial display process wiggles the ears. However, often there is blending across the same set of facial features. For instance, when vocalizing in a "sad" affective state, the control for lip synchronization with facial emphasis competes for the same facial features needed to convey sadness. Here, blending must take place to maintain a consistent expression of affective state.

Figure 10.2 illustrates how the facial feature arbitration is implemented. It is a priority-based scheme, where higher-level subsystems bid for each facial feature that they want to control. The bids are broken down into each observable movement of the facial feature. Instead of bidding for the left ear as a whole, separate bids are made for left ear elevation and left ear rotation. To promote coherency, the bids for each component movement of a facial feature by a given subsystem are generally set to be the same. The flexibility is present to have different subsystems control them independently, should it be appropriate to do so. The highest bid wins the competition and gets to forward its request to the underlying facial feature primitive. The request includes the target position, velocity, and type of movement (postural or rhythmic).

The priorities are defined by hand, although the bid for each facial feature changes dynamically depending on the current motor skill. There are general rules of thumb that are followed. For a low to moderate "emotive" intensity level, the facial expression subsystem sets the expression baseline and has the lowest priority. It is always active when no other facial function is to be performed. The "emotive" baseline can be over-ridden by "voluntary" movements (e.g., facial gestures) as well as behavioral responses (such as "sleeping"). If an emotional response is evoked (due to a highly active emotion process), however, the facial expression will be given a higher priority so that it will be expressed. Lip synchronization has the highest priority over the lips and mouth whenever a request to speak has been made. Thus, whenever the robot says something, the lips and jaw coordinate with the vocal modality.

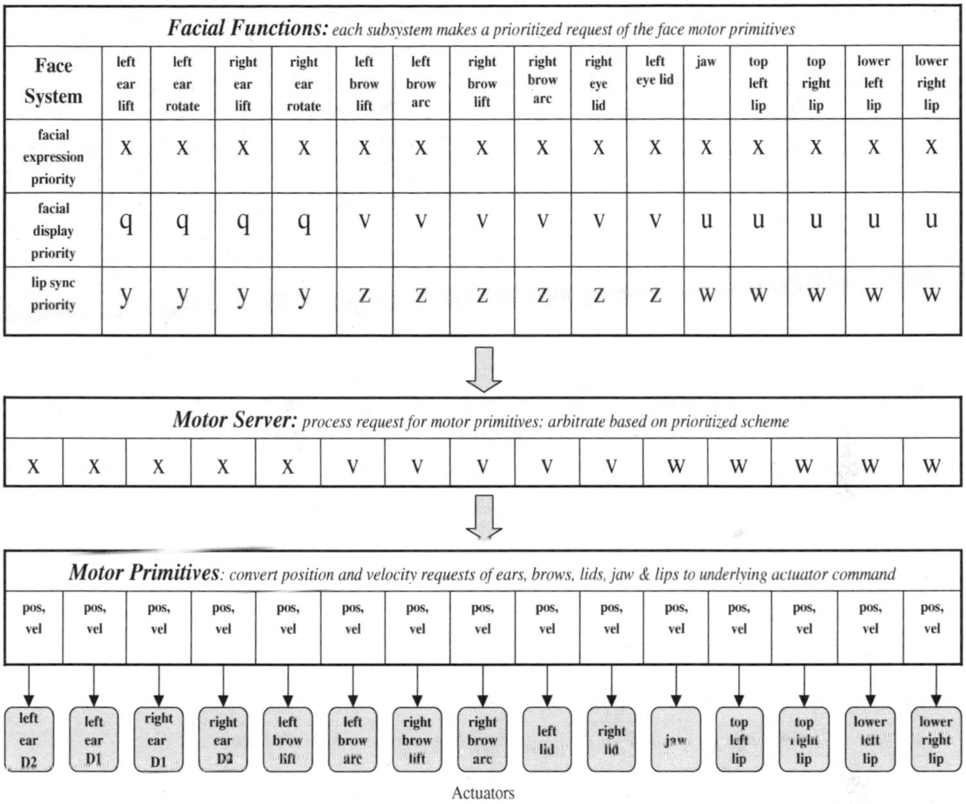

Facial Functions: *each subsystem makes a prioritized request of the face motor primitives*

Face System	left ear lift	left ear rotate	right ear lift	right ear rotate	left brow lift	left brow arc	right brow lift	right brow arc	right eye lid	left eye lid	jaw	top left lip	top right lip	lower left lip	lower right lip
facial expression priority	X	X	X	X	X	X	X	X	X	X	X	X	X	X	X
facial display priority	q	q	q	q	v	v	v	v	v	v	u	u	u	u	u
lip sync priority	y	y	y	y	z	z	z	z	z	z	w	w	w	w	w

Motor Server: *process request for motor primitives; arbitrate based on prioritized scheme*

X	X	X	X	X	V	V	V	V	V	W	W	W	W	W

Motor Primitives: *convert position and velocity requests of ears, brows, lids, jaw & lips to underlying actuator command*

pos, vel	pos, vel	pos, vel	pos, vel	pos, vel	pos, vel	pos, vel	pos, vel	pos, vel	pos, vel	pos, vel	pos, vel	pos, vel	pos, vel	pos, vel

Actuators: left ear D2 | left ear D1 | right ear D1 | right ear D2 | left brow lift | left brow arc | right brow lift | right brow arc | left lid | right lid | jaw | top left lip | top right lip | lower left lip | lower right lip

Figure 10.2
Face arbitration is handled through a dynamic priority scheme. In the figure, q, u, v, w, x, y, z are hand-coded priorities. These are updated whenever a new request is made to a face motor subsystem. The actuators belonging to each type of facial feature are given the same priority so that they serve the same function. At the motor server level, the largest priorities get control of those motors. In this example, the ears shall serve the expression function, the eyebrows shall serve the display function, and the lips shall serve the lip synchronization function.

The facial emphasis component of lip synchronization modulates the facial features about the established baseline. In this way, the rest of the face blends with the underlying facial expression. This is critical for having face, voice, and body all convey a similar emotional state.

The Facial Function Layer

The highest level of the face control hierarchy consists of three subsystems: *emotive facial expression,* communicative *facial display and behavior,* and *lip synchronization and facial emphasis*. Each subsystem serves a different facial function. The *emotive facial expression* subsystem is responsible for generating expressions that convey the robot's current motivational state. Recall that the control of facial displays and behavior was partially covered in chapter 9.

The lip synchronization and facial emphasis system is responsible for coordinating lips, jaw, and the rest of the face with speech. The lips are synchronized with the spoken phonemes as the rest of the face lends coordinated emphasis. See chapter 11 for the details of how Kismet's lip synchronization and facial emphasis system is implemented.

The facial display and behavior subsystem is responsible for postural displays of the face (such as raising the brows at the end of a speaking turn), animated facial gestures (such as exuberantly wiggling the ears in an attention grabbing display), and behavioral responses (such as flinching in response to a threatening stimulus). Taken as a whole, the facial display system encompasses all those facial behaviors not directly generated by the emotional system. Currently, they are modeled as simple routines that are evoked by the motor skills system (as presented in chapter 9) for a specified amount of time and then released (see table 10.1). The motor skills system handles the coordination of these facial

Table 10.1
A summary of Kismet's facial displays.

Stereotyped Display	Description
Sleep and Wake-up Display	Associated with the behavioral response of going to "sleep" and "waking up."
Grimace and Flinch Display	Associated with the fear response. The eyes close, the ears cover and are lowered, the mouth frowns. It is evoked in conjunction with the `flee` behavioral response.
Calling Display	Associated with the calling behavior. It is a stereotyped movement designed to get people's attention and encourage them to approach the robot. The ears waggle exuberantly (causing significant noise), the lips have slight smile. It includes a forward postural shift and head/eye orientation to the person. If the eye-detector can find the eyes, the robot makes eye contact with the person. The robot also vocalizes with an aroused affect. The desired impression is for the targeted person to interpret the display as the robot calling to them.
Greet Display	A stereotyped response involving a smile and small waggling of the ears.
Raise Brows Display	A social cue used to signal the end of the robot's turn in vocal proto-dialog. It is used whenever the robot should look expectant to prompt the human to respond. If the eyes are found, the robot makes eye-contact with the person
Perk Ears Reflex	A social feedback cue whenever the robot hears and sound. It is used as a little acknowledgement that the robot heard the person say something.
Blink Reflex	A social cue often used when the robot has finished its speaking turn. It is often accompanied by a gaze shift away from the listener.
Startle Reflex	A reflex in response to a looming stimulus. The mouth opens, the lips are rounded, the ears perk, the eyes widen, and the eyebrows elevate.

displays with vocal, postural, and gaze/orientation behavior. Ultimately, this subsystem might include learned movements that could be acquired during imitative facial games with the caregiver.

The emotive facial expression subsystem is responsible for generating a facial expression that mirrors the robot's current affective state. This is an important communication signal for the robot. It lends richness to social interactions with humans and increases their level of engagement. For the remainder of this chapter, I describe the implementation of this system in detail. I also discuss how affective postural shifts complement the facial expressions and lend strength to the overall expression. The expressions are analyzed and their readability evaluated by subjects with minimal to no prior familiarity with the robot (Breazeal, 2000a).

10.3   Generation of Facial Expressions

There have been only a few expressive autonomous robots (Velasquez, 1998; Fujita & Kageyama, 1997) and a few expressive humanoid faces (Hara, 1998; Takanobu et al., 1999). The majority of these robots are only capable of a limited set of fixed expressions (a single happy expression, a single sad expression, etc.). This hinders both the believability and readability of their behavior. The expressive behavior of many robotic faces is not life-like (or believable) because of their discrete, mechanical, and reflexive quality—transitioning between expressions like a switch being thrown. This discreteness and discontinuity of transitions limits the readability of the face. It lacks important cues for the intensity of the underlying affective state. It also lacks important cues for the transition dynamics between affective states.

Insights from Animation

Classical and computer animators have a tremendous appreciation for the challenge in creating believable behavior. They also appreciate the role that expressiveness plays in this endeavor. A number of animation guidelines and techniques have been developed for achieving life-like, believable, and compelling animation (Thomas & Johnston, 1981; Parke & Waters, 1996). These rules of thumb explicitly consider audience perception. The rules are designed to create behavior that is rich and interesting, yet easily understandable to the human observer. Because Kismet interacts with humans, the robot's expressive behavior must cater to the perceptual needs of the human observer. This improves the quality of social interaction because the observer feels that she understands the robot's behavior. This helps her to better predict the robot's responses to her, and in turn to shape her own responses to the robot.

Of particular importance is timing: how to sequence and how to transition between actions. A cardinal rule of timing is to *do one thing at a time*. This allows the observer to witness and interpret each action. It is also important that each action last for a sufficiently long time span for the observer to read it. Given these two guidelines, Kismet expresses only one emotion at a time, and each expression has a minimum persistence of several seconds before it decays. The time of intense expression can be extended if the corresponding "emotion" continues to be highly active.

The transitions between expressive behaviors should be smooth. The build-up and decay of expressive behavior can occur at different rates, but it should not be discontinuous like throwing a switch. Animators interpolate between target frames for this purpose, while controlling the morphing rate from the initial posture to the final posture. The physics of Kismet's motors does the smoothing for us to some extent, but the velocities and accelerations between postures are important. An aroused robot will exhibit quick movements of larger amplitude. A subdued robot will move more sluggishly. The accelerations and decelerations into these target postures must also be considered. Robots are often controlled for speed and accuracy—to achieve the fastest response time possible with minimal overshoot. Biological systems don't move like this. For this reason, Kismet's target postures as well as the velocities and accelerations that achieve them are carefully considered.

Animators take a lot of care in drawing the audience's attention to the part of the scene where an important action is about to take place. By doing so, the audience's attention is directed to the right place at the right time so that they do not miss out on important information. To enhance the readability and understandability of Kismet's behavior, its direction of gaze and facial expression serve this purpose. People naturally tend look at what Kismet is looking at. They observe the expression on its face to see how the robot is affectively assessing the stimulus. This helps them to predict the robot's behavior. If the robot looks at a stimulus with an interested expression, the observer predicts that the robot will continue to engage the stimulus. Alternatively, if the robot has a frightened expression, the observer is not surprised to witness a fleeing response soon afterwards. Kismet's expression and gaze precede the behavioral response to make it understandable and predictable to the human who interacts with it.

Expression is not just conveyed through face, but through the entire body. In general, Kismet's expressive shifts in posture may modify the motor commands of more task-based motor skills (such as orienting toward a particular object). Consequently, the issue of expressive blending with neck and eye motors arises. To accomplish successful blending, the affective state determines the default posture of the robot, and the task-based motor commands are treated as offsets from this posture. To add more complexity, the robot's level of arousal sets the velocities and accelerations of the task-based movements. This causes the robot to move sluggishly when arousal is low, and to move in a darting manner when in a high arousal state.

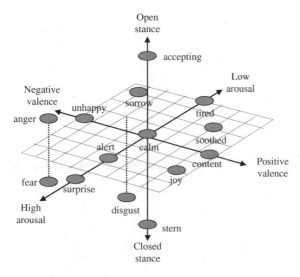

Figure 10.3
The affect space consists of three dimensions. The extremes are: high arousal, low arousal, positive valence, negative valence, open stance, and closed stance. The `emotional` processes can be mapped to this space.

Generating Emotive Expression

Kismet's facial expressions are generated using an interpolation-based technique over a three-dimensional space (see figure 10.3). The three dimensions correspond to arousal, valence, and stance. Recall in chapter 8, the same three attributes are used to affectively assess the myriad of environmental and internal factors that contribute to Kismet's affective state. I call the space defined by the $[A, V, S]$ trio the *affect space*. The current affective state occupies a single point in this space at a time. As the robot's affective state changes, this point moves about within this space. Note that this space not only maps to "emotional" states (e.g., `anger`, `fear`, `sadness`, etc.) but also to the level of arousal as well (e.g., `excitement` and `fatigue`). A range of expressions generated with this technique is shown in figure 10.4. The procedure runs in real-time, which is critical for social interaction.

The affect space can be roughly partitioned into regions that map to each `emotion` process (see figure 10.3). The mapping is defined to be coarse at first, and the emotion system is initially configured so that only limited regions of the overall space are frequented often. The intention was to support the possibility of "emotional" and expressive development, where the `emotion` processes continue to refine as secondary "emotions" are acquired through experience and associated with particular regions in affect space with their corresponding facial expressions.

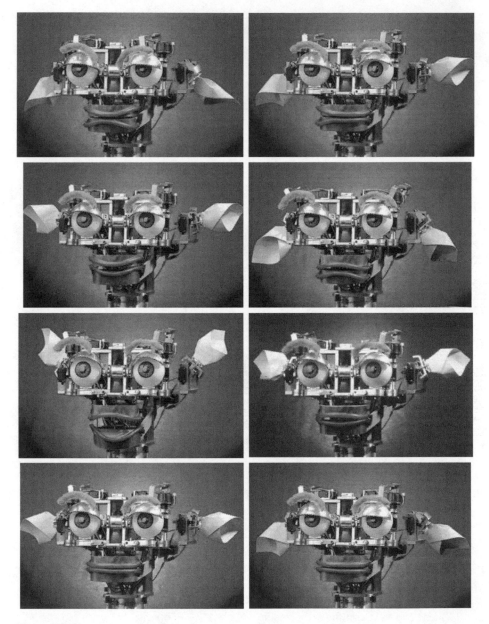

Figure 10.4
Kismet is capable of generating a continuous range of expressions of various intensities by blending the basis
facial postures. Facial movements correspond to affect dimensions in a principled way. A sampling is shown here.
These can also be viewed, with accompanying vocalizations, on the included CD-ROM.

There are nine *basis* (or *prototype*) *postures* that collectively span this space of emotive expressions. Although some of these postures adjust specific facial features more strongly than the others, each prototype influences most if not all of the facial features to some degree. For instance, the valence prototypes have the strongest influence on lip curvature, but can also adjust the positions of the ears, eyelids, eyebrows, and jaw. The basis set of facial postures has been designed so that a specific location in affect space specifies the relative contributions of the prototype postures in order to produce a net facial expression that faithfully corresponds to the active emotion. With this scheme, Kismet displays expressions that intuitively map to the human emotions of anger, disgust, fear, happiness, sorrow, and surprise. Different levels of arousal can be expressed as well from interest, to calm, to weariness.

There are several advantages to generating the robot's facial expression from this affect space. First, this technique allows the robot's facial expression to reflect the nuance of the underlying assessment. Even through there is a discrete number of emotion processes, the expressive behavior spans a continuous space. Second, it lends clarity to the facial expression since the robot can only be in a single affective state at a time (by choice) and hence can only express a single state at a time. Third, the robot's internal dynamics are designed to promote smooth trajectories through affect space. This gives the observer a lot of information about how the robot's affective state is changing, which makes the robot's facial behavior more interesting. Furthermore, by having the face mirror this trajectory, the observer has immediate feedback as to how their behavior is influencing the robot's internal state. For instance, if the robot has a distressed expression upon its face, it may prompt the observer to speak in a soothing manner to Kismet. The soothing speech is assimilated into the emotion system where it causes a smooth decrease in the arousal dimension and a push toward slightly positive valence. Thus, as the person speaks in a comforting manner, it is possible to witness a smooth transition to a subdued expression. However, if the face appeared to grow more aroused, then the person may stop trying to comfort the robot verbally and perhaps try to please the robot by showing it a colorful toy.

The six *primary prototype postures* sit at the extremes of each dimension (see figure 10.5). They correspond to high arousal, low arousal, negative valence, positive valence, open (approaching) stance, and closed (withdrawing) stance. The high arousal prototype, P_{high}, maps to the expression for surprise. The low arousal prototype, P_{low}, corresponds to the expression for fatigue (note that sleep is a behavioral response, so it is covered in the facial display subsystem). The positive valence prototype, $P_{positive}$, maps to a content expression. The negative valence prototype, $P_{negative}$, resembles an unhappy expression. The closed stance prototype, P_{closed}, resembles a stern expression, and the open stance prototype, P_{open}, resembles an accepting expression.

The three affect dimensions also map to affective postures. There are six basis postures defined which span the space. High arousal corresponds to an erect posture with a slight

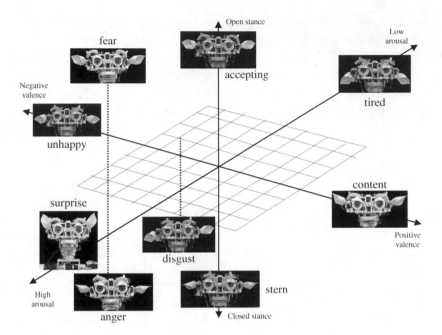

Figure 10.5
This diagram illustrates where the basis postures are located in affect space.

upward chin. Low arousal corresponds to a slouching posture where the neck lean and head tilt are lowered. The posture remains neutral over the valence dimension. An open stance corresponds to a forward lean movement, which suggests strong interest toward the stimuli the robot is leaning toward. A closed stance corresponds to withdraw, reminiscent of shrinking away from whatever the robot is looking at. In contrast to the facial expressions (which are continually expressed), the affective postures are only expressed when the corresponding emotion process has sufficiently strong activity. When expressed, the posture is held for a minimum period of time so that the observer can read it, and then it is released. The facial expression, of course, remains active. The posture is presented for strong conveyance of a particular affective state.

The remaining three facial prototypes are used to strongly distinguish the expressions for disgust, anger, and fear. Recall that four of the six primary `emotions` are characterized by negative valence. Whereas the primary six basis postures (presented above) can generate a range of negative expressions from distress to sadness, the expressions for intense anger (rage), intense fear (terror), and intense disgust have some uniquely distinguishing features. For instance, the prototype for disgust, $P_{disgust}$, is unique in its asymmetry (typical of this expression). The prototypes for anger, P_{anger}, and fear, P_{fear}, each have a distinct configuration for the lips (furious lips form a snarl, terrified lips form a grimace).

Each dimension of the affect space is bounded by the minimum and maximum allowable values of $(min, max) = (-1250, 1250)$. The placement of the prototype postures is given in figure 10.5. The current net affective assessment from the emotion system defines the $[A, V, S] = (a, v, s)$ point in affect space. The specific (a, v, s) values are used to weight the relative motor contributions of the basis postures. Using a weighted interpolation scheme, the net emotive expression, P_{net}, is computed. The contributions are computed as follows:

$$P_{net} = C_{arousal} + C_{valence} + C_{stance} \tag{10.1}$$

where
P_{net} is the emotive expression computed by weighted interpolation
$C_{arousal}$ is the weighted motor contribution due to the arousal state
$C_{valence}$ is the weighted motor contribution due to the valence state
C_{stance} is the weighted motor contribution due to stance state

These contributions are specified by the equations:

$$C_{arousal} = \alpha P_{high} + (1 - \alpha) P_{low}$$
$$C_{valence} = \beta P_{positive} + (1 - \beta) P_{negative}$$
$$C_{stance} = F(a, v, s, n) + (1 - \delta)(\gamma P_{open} + (1 - \gamma) P_{closed})$$

where the fractional interpolation coefficients are:
$\alpha, 0 \leq \alpha \leq 1$ for arousal
$\beta, 0 \leq \beta \leq 1$ for valence
$\gamma, 0 \leq \gamma \leq 1$ for stance
$\delta, 0 \leq \delta \leq 1$ for the specialized prototype postures

such that δ and $F(A, V, S, N)$ are defined as follows:
$$\delta = f_{anger}(A, V, S, N) + f_{fear}(A, V, S, N) + f_{disgust}(A, V, S, N)$$
$$\begin{aligned} F(A, V, S, N) = &\, f_{anger}(A, V, S, N) \cdot P_{anger} + \\ &\, f_{fear}(A, V, S, N) \cdot P_{fear} + \\ &\, f_{disgust}(A, V, S, N) \cdot P_{disgust} \end{aligned}$$

The weighting function $f_i(A, V, S, N)$ limits the influence of each specialized prototype posture to remain local to their region of affect space. Recall, there are three specialized postures, P_i, for the expressions of anger, fear, and disgust. Each is located at $(A_{P_i}, V_{P_i}, S_{P_i})$ where A_{P_i} corresponds to the arousal coordinate for posture P_i, V_{P_i} corresponds to the valence coordinate, and S_{P_i} corresponds to the stance coordinate. Given the current net affective state (a, v, s) as computed by the emotion system, one can compute the displacement from (a, v, s) to each $(A_{P_i}, V_{P_i}, S_{P_i})$. For each P_i, the weighting function $f_i(A, V, S, N)$ decays linearly with distance from $(A_{P_i}, V_{P_i}, S_{P_i})$. The weight is bounded between $0 \leq f_i(A, V, S, N) \leq 1$, where the maximum value occurs at $(A_{P_i}, V_{P_i}, S_{P_i})$. The argument N defines the radius of influence which is kept fairly small so that the contribution for each specialized prototype posture does not overlap with the others.

Comparison to Componential Approaches

It is interesting to note the similarity of this scheme with the affect dimensions viewpoint of emotion (Russell, 1997; Smith & Scott, 1997). Instead of viewing emotions in terms of categories (happiness, anger, fear, etc.), this viewpoint conceptualizes the dimensions that could span the relationship between different emotions (arousal and valence, for instance). Instead of taking a production-based approach to facial expression (how do emotions generate facial expressions), Russell (1997) takes a perceptual stance (what information can an observer read from a facial expression). For the purposes of Kismet, this perspective makes a lot of sense, given the issue of readability and understandability.

Psychologists of this view posit that facial expressions have a systematic, coherent, and meaningful structure that can be mapped to affective dimensions (Russell, 1997; Lazarus, 1991; Plutchik, 1984; Smith, 1989; Woodworth, 1938). (See figure 10.6 for an example.) Hence, by considering the individual facial action components that contribute to that structure, it is possible to reveal much about the underlying properties of the emotion being expressed. It follows that some of the individual features of expression have inherent signal value. This promotes a signaling system that is robust, flexible, and resilient (Smith & Scott, 1997). It allows for the mixing of these components to convey a wide range of affective messages, instead of being restricted to a fixed pattern for each emotion. This variation allows fine-tuning of the expression, as features can be emphasized, de-emphasized, added, or omitted as appropriate. Furthermore, it is well-accepted that any emotion can be conveyed equally well by a range of expressions, as long as those expressions share a family resemblance. The resemblance exists because the expressions share common facial action units. It is also known that different expressions for different emotions share some of the same face action components (the raised brows of fear and surprise, for instance). It is hypothesized by Smith and Scott that those features held in common assign a shared affective meaning to each facial expression. The raised brows, for instance, convey attentional activity for both fear and surprise.

Russell (1997) argues the human observer perceives two broad affective categories on the face, arousal and pleasantness. As shown in figure 10.6, Russell maps several emotions and corresponding expressions to these two dimensions. This scheme, however, seems fairly limiting for Kismet. First, it is not clear how all the primary emotions are represented with this scheme (disgust is not accounted for). It also does not account for positively valenced

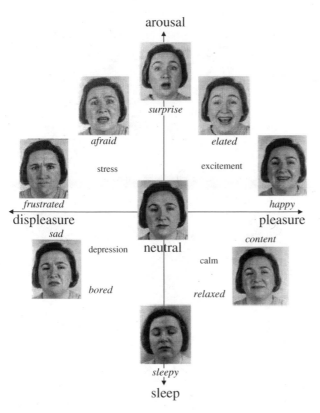

Figure 10.6
Russell's pleasure-arousal space for facial expression.

yet reserved expressions such as a coy smile or a sly grin (which hint at a behavioral bias to withdraw). More importantly, *anger* and *fear* reside in very close proximity to each other despite their very different behavioral correlates. From an evolutionary perspective, the behavioral correlate of anger is to attack (which is a very strong approaching behavior), and the behavioral correlate for fear is to escape (which is a very strong withdrawing behavior). These are stereotypical responses derived from cross-species studies—obviously human behavior can vary widely. Nonetheless, from a practical engineering perspective of *generating* expression, it is better to separate these two emotional responses by a greater distance to minimize accidental activation of one instead of the other. Adding the stance dimension addressed these issues for Kismet.

Given this three dimensional affect space, this approach resonates well with the work of Smith and Scott (1997). They posit a three dimensional space of *pleasure-displeasure* (maps

Table 10.2
A possible mapping of facial movements to affective dimensions proposed by Smith and Scott (1997). An up arrow indicates that the facial action is hypothesized to increase with increasing levels of the affective meaning dimension. A down arrow indicates that the facial action increases as the affective meaning dimension decreases. For instance, the lip corners turn upwards as "pleasantness" increases, and lower with increasing "unpleasantness."

Meaning	Facial Action							
	Eyebrow Frown	Raise Eyebrows	Raise upper Eyelid	Raise Lower Eyelid	Up Turn Lip Corners	Open Mouth	Tighten Mouth	Raise Chin
Pleasantness	⬇				⬆	⬆	⇩	⬇
Goal Obstacle/Discrepancy Anticipated Effort	⬆ ⇧							
Attentional Activity		⬆	⬆					
Certainty		⇩		⇧		⇧		
Novelty		⇧	⇧					
Personal Agency/Control		⇩	⇩			⇩		

to valence here), *attentional activity* (maps to arousal here), and *personal agency/control* (roughly maps to stance here). Table 10.2 summarizes their proposed mapping of facial actions to these dimensions. They posit a fourth dimension that relates to the intensity of the expression. For Kismet, the expressions become more intense as the affect state moves to more extreme values in the affect space. As positive valence increases, Kismet's lips turn upward, the mouth opens, and the eyebrows relax. However, as valence decreases, the brows furrow, the jaw closes, and the lips turn downward. Along the arousal dimension, the ears perk, the eyes widen, brows elevate and the mouth opens as arousal increases. Along the stance dimension, increasing positive values cause the eyebrows to arc outwards, the mouth to open, the ears to open, and the eyes to widen. These face actions roughly correspond to a decrease in personal agency/control in Smith and Scott's framework. For Kismet, it engenders an expression that looks more eager and accepting (or more uncertain for negative emotions). Although Kismet's dimensions do not map exactly to those hypothesized by Smith and Scott, the idea of combining meaningful facial movements in a principled manner to span the space of facial expressions, and to also relate them in a consistent way to emotion categories, holds strong.

10.4   Analysis of Facial Expressions

Ekman and Friesen (1982) developed a commonly used facial measurement system called *FACS*. The system measures the face itself as opposed to trying to infer the underlying emotion given a particular facial configuration. This is a comprehensive system that distinguishes all possible *visually* distinguishable facial movements. Every such facial movement is the result of muscle action (see figure 10.7 and table 10.3). The earliest work in this area dates back to Duchenne (1806–1875), one of the first anatomists to explore how facial muscles change the appearance of the face (Duchenne, 1990). Based on a deep understanding of how muscle contraction changes visible appearance, it is possible to decompose any facial movement into anatomically minimal action units. FACS has defined 33 distinct *action units* for the human face, many of which use a single muscle. It is possible for up to two or

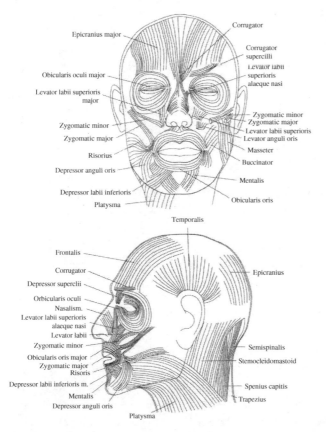

Figure 10.7
A schematic of the muscles of the face. Front and side views from Parke and Waters (1996).

Table 10.3
A summary of how FACS action units and facial muscles map to facial expressions for the primary emotions. Adapted from Smith and Scott (1997).

					Facial Action			
	Eyebrow Frown	Raise Eyebrows	Raise upper Eyelid	Raise Lower Eyelid	Up Turn Lip Corners	Down Turn Lip Corners	Open Mouth	Raise Upper Lip
Muscular Basis	corrugator supercilii	medial frontalis	levator palpebrae superioris	orbicularis oculi	zygomaticus major	depressor anguli oris	orbicularis oris	levator labii superioris
Action units	4	1	5	6,7	12	15	26,27	9,10

					Emotion Expressed			
Happiness				X	X		X	
Surprise		X	X				X	
Anger	X		X	X				
Disgust	X			X				X
Fear	X	X	X				X	
Sadness	X	X				X		

three muscles to map to a given action unit, since facial muscles often work in concert to adjust the location of facial features, and to gather, pouch, bulge, or wrinkle the skin.

To analyze Kismet's facial expressions, FACS can be used as a guideline. This must obviously be done within reason as Kismet lacks many of the facial features of humans (most notably, skin, teeth, and nose). The movements of Kismet's facial mechanisms, however, were designed to roughly mimic those changes that arise in the human face due to the contraction of facial muscles. Kismet's eyebrow movements are shown in figure 10.8, and the eyelid movements in figure 10.9. Kismet's ears are primarily used to convey arousal and stance as shown in figure 10.10. The lip and jaw movements are shown in figure 10.11.

Using the FACS system, Smith and Scott (1997) have compiled mappings of FACS action units to the expressions corresponding to anger, fear, happiness, surprise, disgust, and sadness based on the observations of others (Darwin, 1872; Frijda, 1969; Scherer, 1984; Smith, 1989). Table 10.3 associates an action unit with an expression if two or more of these sources agreed on the association. The facial muscles employed are also listed. Note that these are not inflexible mappings. Any emotion can be expressed by a family of expressions, and the expressions vary in intensity. Nonetheless, this table highlights several key features.

Of the seven action units listed in the table, Kismet lacks only one (the lower eyelid). Of the facial features it does possess, it is capable of all the independent movements listed (given its own idiosyncratic mechanics). Kismet performs some of these movements in a manner that is different, yet roughly analogous, to that of a human. The series of figures,

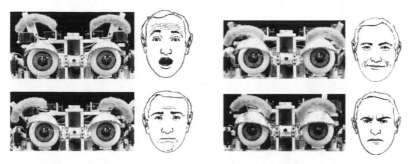

Figure 10.8
Kismet's eyebrow movements for expression. To the right, there is a human sketch displaying the corresponding eyebrow movement (Faigin, 1990). From top to bottom they are surprise, uncertainty or sorrow, neutral, and anger. The eyelids are also shown to lower as one moves from the top left figure to the bottom right figure.

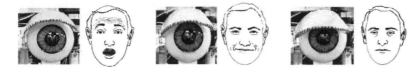

Figure 10.9
Kismet's eyelid movements for expression. To the right of each image of Kismet's eye, there is a human sketch displaying an analogous eyelid position (Faigin, 1990). Kismet's eyelid rests just above the pupil for low arousal states. It rests just below the iris for neutral arousal states. It rests above the iris for high arousal states.

figures 10.8 to 10.11, relates the movement of Kismet's facial features to those of humans. (Video demonstrations of these movements can also be seen on the included CD-ROM.) There are two notable discrepancies. First, the use of the eyelids in Kismet's angry expression differs. In conjunction with brow knitting, Kismet lowers its eyelids to simulate a squint that is accomplished by raising both the lower and upper eyelids in humans. The second is the manner of arcing the eyebrows away from the centerline to simulate the brow configuration in sadness and fear. For humans, this corresponds to simultaneously knitting and raising the eyebrows. See figure 10.8.

Overall, Kismet does address each of the facial movements specified in the table (save those requiring a lower eyelid) in its own peculiar way. One can ask the questions: *How do people identify Kismet's facial expressions with human expressions?*, and *Do they map Kismet's distinctive facial movements to the corresponding human counterparts?*

Comparison with Line Drawings of Human Expressions

To explore these questions, I asked naive subjects to perform a comparison task where they compared color images of Kismet's expressions with a series of line drawings of human

elevated

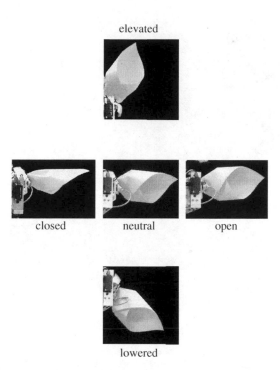

closed   neutral   open

lowered

Figure 10.10
Kismet's ear movements for expression. There is no human counterpart, but they move somewhat like that of an animal. They are used to convey arousal by either pointing upward as shown in the upper figure, or by pointing downward as shown in the bottom figure. The ears also convey approach (the ears rotate forward as shown to the right) versus withdraw (the ears close as shown to the left). The central figure shows the ear in the neutral position.

expressions. It seemed unreasonable to have people compare images of Kismet with human photos since the robot lacks skin. However, the line drawings provide a nice middle ground. The artist can draw lines that suggest the wrinkling of skin, but for the most part this is minimally done. We used a set of line drawings from (Faigin, 1990) to do the study.

Ten subjects filled out the questionnaire. Five of the subjects were children (11 to 12 years old), and five were adults (ranging in age from 18 to 50). The gender split was four females and six males. The adults had never seen Kismet before. Some of the children reported having seen a short school magazine article, so had minimal familiarity.

The questionnaire was nine pages long. On each page was a color image of Kismet in one of nine facial expressions (from top to bottom, left to right they correspond to anger, disgust, happiness, content, surprise, sorrow, fear, stern, and a sly grin). Adjacent to the robot's picture was a set of twelve line drawings labeled *a* though *l*. The drawings are shown in figure 10.12 with my emotive labels. The subject was asked to circle the line

Figure 10.11
Kismet's lip movements for expression. Alongside each of Kismet's lip postures is a human sketch displaying an analogous posture (Faigin, 1990). On the left, top to bottom, are: disgust, fear, and a frown. On the right, top to bottom, are: surprise, anger, and a smile.

drawing that most closely resembled the robot's expression. There was a short sequence of questions to probe the similarity of the robot to the chosen line drawing. One question asked how similar the robot's expression was to the selected line drawing. Another question asked the subject to list the labels of any other drawings they found to resemble the robot's expression and why. Finally, the subject could write any additional comments on the sheet. Table 10.4 presents the compiled results.

The results are substantially above random chance (8 percent), with the expressions corresponding to the primary emotions giving the strongest performance (70 percent and above). Subjects could infer the intensity of expression for the robot's expression of happiness (a contented smile versus a big grin). They had decent performance (60 percent) in matching Kismet's stern expression (produced by zero arousal, zero valence, and strong negative stance). The "sly grin" is a complex blend of positive valence, neutral arousal, and closed stance. This expression gave the subjects the most trouble, but their matching performance is still significantly above chance.

The misclassifications seem to arise from three sources. Certain subjects were confused by Kismet's lip mechanics. When the lips curve either up or down, there is a slight curvature in the opposite direction at the lever arm insertion point. Most subjects ignored the bit of

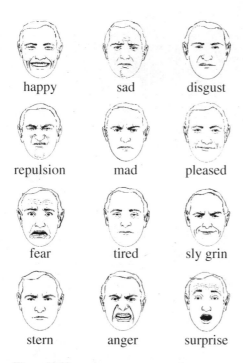

happy   sad   disgust

repulsion   mad   pleased

fear   tired   sly grin

stern   anger   surprise

Figure 10.12
The sketches used in the evaluation, adapted from Faigin (1990). The labels are for presentation purposes here; in the study they were labeled with the letters ranging from *a* through *l*.

curvature at the extremes of the lips, but others tried to match it to the lips in the line drawings. Occasionally, Kismet's frightened grimace was matched to a smile, or its smile matched to repulsion. Some misclassifications arose from matching the robot's expression to a line drawing that conveyed the same sentiment to the subject. For instance, Kismet's expression for disgust was matched to the line sketch of the "sly grin" because the subject interpreted both as "sneering" although none of the facial features match. Some associated Kismet's surprise expression with the line drawing of "happiness." There seems to be a positive valence communicated though Kismet's expression for surprise. Misclassifications also arose when subjects only seemed to match a single facial feature to a line drawing instead of multiple features. For instance, one subject matched Kismet's stern expression to the sketch of the "sly grin," noting the similarity in the brows (although the robot is not smiling). Overall, the subjects seem to intuitively match Kismet's facial features to those of the line drawings, and interpreted their shape in a similar manner. It is interesting to note that the robot's ears seem to communicate an intuitive sense of arousal to the subjects as well.

Table 10.4
Human subject's ability to map Kismet's facial features to those of a human sketch. The human sketches are shown in figure 10.12. An intensity difference was explored (content versus happy). An interesting blend of positive valence with closed stance was also tested (the sly grin).

	most similar sketch	data	comments
anger	anger	10/10	Shape of mouth and eyebrows are strongest reported cues
disgust	disgust	8/10	Shape of mouth is strongest reported cue
	sly grin	2/10	Described as "sneering"
fear	fear	7/10	Shape of mouth and eyes are strongest reported cues; mouth open "aghast"
	surprise	1/10	Subject associates look of "shock" with sketch of "surprise" over "fear"
	happy	1/10	Lip mechanics cause lips to turn up at ends, sometimes confused with a weak smile
joy	happy	7/10	Report lips and eyes are strongest cues; ears may provide arousal cue to lend intensity
	content	1/10	Report lips used as strongest cue
	repulsion	1/10	Lip mechanics turn lips up at end, causing shape reminiscent of lips in repulsion sketch
	surprise	1/10	Perked ears, wide eyes lend high arousal; sometimes associated with a pleasant surprise
sorrow	sad	9/10	Lips reported as strongest cue, low ears may lend to low arousal
	repulsion	1/10	Lip mechanics turn lips up and end, causing shape reminiscent of repulsion sketch
surprise	surprise	9/10	Reported open mouth, raised brows, wide eyes and elevated ears all lend to high arousal
	happy	1/10	Subject remarks on similarity of eyes, but not mouth
pleased	content	9/10	Reported relaxed smile, ears, and eyes lend low arousal and positive valence
	sly grin	1/10	Subject reports the robot exhibiting a reserved pleasure; associated with the "sly grin" sketch
sly grin	sly grin	5/10	Lips and eyebrows reported as strongest cues
	content	3/10	Subjects use robot's grin as the primary cue
	stern	1/10	Subject reports the robot looking "serious" which is associated with "sly grin" sketch
	repulsion	1/10	Lip mechanics curve lips up at end; subject sees similarity with lips in "repulsion" sketch
stern	stern	6/10	Lips and eyebrows are reported as strongest cues
	mad	1/10	Subject reports robot looking "slightly cross;" cue on robot's eyebrows and pressed lips
	tired	2/10	Subjects may cue in on robot's pressed lips, low ears, lowered eyelids
	sly grin	1/10	Subject reports similarity in brows

10.5 Evaluation of Expressive Behavior

The line drawing study did not ask the subjects what they thought the robot was expressing. Clearly, however, this is an important question for my purposes. To explore this issue, a separate questionnaire was devised. Given the wide variation in language that people use to describe expressions and the small number of subjects, a forced choice paradigm was adopted.

Seventeen subjects filled out the questionnaire. Most of the subjects were children 12 years of age (note that Kolb et al. [1992] found that the ability to recognize expressions continues to develop, reaching adult level competence at approximately 14 years of age). There were six girls, six boys, three adult men, and two adult women. Again, none of the adults had seen the robot before. Some of the children reported minimal familiarity through reading a children's magazine article. There were seven pages in the questionnaire. Each page had a large color image of Kismet displaying one of seven expressions (anger, disgust, fear, happiness, sorrow, surprise, and a stern expression). The subjects could choose the best match from ten possible labels (accepting, anger, bored, disgust, fear, joy, interest, sorrow, stern, surprise). In a follow-up question, they could circle any other labels that they thought could also apply. With respect to their best-choice answer, they were asked to specify on a ten-point scale how confident they were of their answer, and how intense they found the expression. The complied results are shown in table 10.5. The subjects' responses were significantly above random choice (10 percent), ranging from 47 percent to 83 percent.

Some of the misclassifications are initially confusing, but made understandable in light of the aforementioned study. Given that Kismet's surprise expression seems to convey positive valence, it is not surprising that some subjects matched it to *joy*. The knitting of the brow in Kismet's stern expression is most likely responsible for the associations with negative emotions such as anger and sorrow. Often, negatively valenced expressions were

Table 10.5
This table summarizes the results of the color-image-based evaluation. The questionnaire was forced choice where the subject chose the emotive word that best matched the picture.

	accepting	anger	bored	disgust	fear	joy	interest	sorrow	stern	surprise	% correct
anger	5.9	76.5	0	0	5.9	11.7	0	0	0	0	76.5
disgust	0	17.6	0	70.6	5.9	0	0	0	5.9	0	70.6
fear	5.9	5.9	0	0	47.1	17.6	5.9	0	0	17.6	47.1
joy	11.7	0	5.9	0	0	82.4	0	0	0	0	82.4
sorrow	0	5.9	0	0	11.7	0	0	83.4	0	0	83.4
stern	7.7	15.4	0	7.7	0	0	0	15.4	53.8	0	53.8
surprise	0	0	0	0	0	17.6	0	0	0	82.4	82.4

Forced-Choice Percentage (random = 10%)

misclassified with negatively valenced labels. For instance, labeling the sad expression with fear, or the disgust expression with anger or fear. Kismet's expression for fear seems to give people the most difficulty. The lip mechanics probably account for the association with joy. The wide eyes, elevated brows, and elevated ears suggest high arousal. This may account for the confusion with surprise.

The still image and line drawing studies were useful in understanding how people read Kismet's facial expressions, but it says very little about expressive posturing. Humans and animals not only express with their face, but with their entire body. To explore this issue for Kismet, I showed a small group of subjects a set of video clips.

There were seven people who filled out the questionnaire. Six were children of age 12, four boys and two girls. One was an adult female. In each clip Kismet performs a coordinated expression using face and body posture. There were seven videos in all (anger, disgust, fear, joy, interest, sorrow, and surprise). Using a forced-choice paradigm, for each video the subject was asked to select a word that best described the robot's expression (anger, disgust, fear, joy, interest, sorrow, or surprise). On a ten-point scale, the subjects were also asked to rate the intensity of the robot's expression and the certainty of their answer. They were also asked to write down any comments they had. The results are compiled in table 10.6. Random chance is 14 percent.

The subjects performed significantly above chance, with overall stronger recognition performance than on the still images alone. The video segments for the expressions of anger, disgust, fear, and sorrow were correctly classified with a higher percentage than the still images. However, there were substantially fewer subjects who participated in the video evaluation than the still image evaluation. The recognition of joy most likely dipped from the still-image counterpart because it was sometimes confused with the expression of interest in the video study. The perked ears, attentive eyes, and smile give the robot a sense of expectation that could be interpreted as interest.

Table 10.6
This table summarizes the results of the video evaluation.

	anger	disgust	fear	joy	interest	sorrow	surprise	% correct
anger	86	0	0	14	0	0	0	86
disgust	0	86	0	0	0	14	0	86
fear	0	0	86	0	0	0	14	86
joy	0	0	0	57	28	0	15	57
interest	0	0	0	0	71	0	29	71
sorrow	14	0	0	0	0	86	0	86
surprise	0	0	29	0	0	0	71	71

Forced-Choice Percentage (random = 14%)

Misclassifications are strongly correlated with expressions having similar facial or postural components. Surprise was sometimes confused for fear; both have a quick withdraw postural shift (the fearful withdraw is more of a cowering movement whereas the surprise posture has more of an erect quality) with wide eyes and elevated ears. Surprise was sometimes confused with interest. Both have an alert and attentive quality, but interest is an approaching movement whereas surprise is more of a startled movement. Sorrow was sometimes confused with disgust; both are negative expressions with a downward component to the posture. The sorrow posture shift is more down and "sagging," whereas the disgust is a slow "shrinking" retreat.

Overall, the data gathered from these small evaluations suggest that people with little to no familiarity with the robot are able to interpret the robot's facial expressions and affective posturing. For this data set, there was no clear distinction in recognition performance between adults versus children, or males versus females. The subjects intuitively correlate Kismet's face with human likenesses (i.e., the line drawings). They map the expressions to corresponding emotion labels with reasonable consistency, and many of the errors can be explained through similarity in facial features or similarity in affective assessment (e.g., shared aspects of arousal or valence).

The data from the video studies suggest that witnessing the movement of the robot's face and body strengthens the recognition of the expression. More subjects must be tested, however, to strengthen this claim. Nonetheless, observations from other interaction studies discussed throughout this book support this hypothesis. For instance, the postural shifts during the affective intent studies (see chapter 7) beautifully illustrate how subjects read and affectively respond to the robot's expressive posturing and facial expression. This is also illustrated in the social amplification studies of chapter 12. Based on the robot's withdraw and approach posturing, the subjects adapt their behavior to accommodate the robot.

10.6   Limitations and Extensions

More extensive studies need to be performed for us to make any strong claims about how accurately Kismet's expressions mirror those of humans. However, given the small sample size, the data suggest that Kismet's expressions are readable by people with minimal to no prior familiarity with the robot.

The evaluations have provided us with some useful input for how to improve the strength and clarity of Kismet's expressions. A lower eyelid should be added. Several subjects commented on this being a problem for them. The FACS system asserts that the movement of the lower eyelid is a key facial feature in expressing the basic emotions. The eyebrow mechanics could be improved. They should be able to elevate at both corners of the brow, as opposed to the arc of the current implementation. This would allow us to more accurately portray the brow movements for fear and sorrow. Kismet's mechanics attempt to approximate this, but the movement could be strengthened. The insertion point of the motor lever arm to the lips needs to be improved, or at least masked from plain view. Several subjects confused the additional curve at the ends for other lip shapes.

In this chapter, I have only evaluated the readability of Kismet's facial expressions. The evaluation of Kismet's facial displays will be addressed in chapter 12 and chapter 13, when I discuss social interactions between human subjects and Kismet.

As a longer term extension, Kismet should be able to exert "voluntary" control over its facial expressions and be able to learn new facial displays. I have a strong interest in exploring facial imitation in the context of imitative games. Certain forms of facial imitation appear very early in human infants (Meltzoff & Moore, 1977). Meltzoff posits that imitation is an important discovery procedure for learning about and understanding people. It may even play a role in the acquisition of a theory of mind. For adult-level human social intelligence, the question of how a robot could have a genuine theory of mind will need to be addressed.

10.7   Summary

A framework to control the facial movements of Kismet has been developed. The expressions and displays are generated in real-time and serve four facial functions. The lip synchronization and facial emphasis subsystem is responsible for moving the lips and face to accompany expressive speech. The emotive facial expression subsystem is responsible for computing an appropriate emotive display. The facial display and behavior subsystem produces facial movements that serve communicative functions (such as regulating turn taking) as well as producing the facial component of behavioral responses. With so many facial functions competing for the face actuators, a dynamic prioritizing scheme was developed. This system addresses the issues of blending as well as sequencing the concurrent requests made by each of the face subsystems. The overall face control system produces facial movements that are timely, coherent, intuitive and appropriate. It is organized in a principled manner so that incremental improvements and additions can be made. An intriguing extension is to learn new facial behaviors through imitative games with the caregiver, as well as to learn their social significance.

11 Expressive Vocalization System

In the very first instance, he is learning that there is such a thing as language at all, that vocal sounds are functional in character. He is learning that the articulatory resources with which he is endowed can be put to the service of certain functions in his own life. For a child, using his voice is doing something; it is a form of action, and one which soon develops its own patterns and its own significant contexts.
—M.A.K. Halliday (1979, p. 10)

From Kismet's inception, the synthetic nervous system has been designed with an eye toward exploring the acquisition of meaningful communication. As Haliday argues, this process is driven internally through motivations and externally through social engagement with caregivers. Much of Kismet's social interaction with its caregivers is based on vocal exchanges when in face-to-face contact. At some point, these exchanges could be ritualized into a variety of vocal games that could ultimately serve as learning episodes for the acquisition of shared meanings. Towards this goal, this chapter focuses on Kismet's vocal production, expression, and delivery. The design issues are outlined below:

Production of novel utterances   Given the goal of acquiring a proto-language, Kismet must be able to experiment with its vocalizations to explore their effects on the caregiver's behavior. Hence the vocalization system must support this exploratory process. At the very least the system should support the generation of short strings of phonemes, modulated by pitch, duration, and energy. Human infants play with the same elements (and more) when exploring their own vocalization abilities and the effect these vocalizations have on their social world.

Expressive speech   Kismet's vocalizations should also convey the affective state of the robot. This provides the caregiver with important information as to how to appropriately engage Kismet. The robot could then use its emotive vocalizations to convey disapproval, frustration, disappointment, attentiveness, or playfulness. As for human infants, this ability is important for meaningful social exchanges with Kismet. It helps the caregiver to correctly read the robot and to treat the robot as an intentional creature. This fosters richer and sustained social interaction, and helps to maintain the person's interest as well as that of the robot.

Lip synchronization   For a compelling verbal exchange, it is also important for Kismet to accompany its expressive speech with appropriate motor movements of the lips, jaw, and face. The ability to lip synchronize with speech strengthens the perception of Kismet as a social creature that expresses itself vocally. A disembodied voice would be a detriment to the life-like quality of interaction that I and my colleagues have worked so hard to achieve in many different ways. Furthermore, it is well-accepted that facial expressions (related to affect) and facial displays (which serve a communication function) are important for verbal communication. Synchronized movements of the face with voice both complement as well as supplement the information transmitted through the verbal channel. For Kismet, the information communicated to the human is grounded in affect. The facial displays are used to help regulate the dynamics of the exchange. (Video demonstrations of Kismet's expressive displays and the accompanying vocalizations are included on the CD-ROM in the second section, "Readable Expressions.")

11.1   Emotion in Human Speech

There has been an increasing amount of work in identifying those acoustic features that vary with the speaker's affective state (Murray & Arnott, 1993). Changes in the speaker's autonomic nervous system can account for some of the most significant changes, where the sympathetic and parasympathetic subsystems regulate arousal in opposition. For instance, when a subject is in a state of fear, anger, or joy, the sympathetic nervous system is aroused. This induces an increased heart rate, higher blood pressure, changes in depth of respiratory movements, greater sub-glottal pressure, dryness of the mouth, and occasional muscle tremor. The resulting speech is faster, louder, and more precisely enunciated with strong high-frequency energy, a higher average pitch, and wider pitch range. In contrast, when a subject is tired, bored, or sad, the parasympathetic nervous system is more active. This causes a decreased heart rate, lower blood pressure, and increased salivation. The resulting speech is typically slower, lower-pitched, more slurred, and with little high frequency energy. Picard (1997) presents a nice overview of work in this area.

Table 11.1 summarizes the effects of emotion in speech tend to alter the pitch, timing, voice quality, and articulation of the speech signal. Several of these features, however, are also modulated by the prosodic effects that the speaker uses to communicate grammatical structure and lexical correlates. These tend to have a more localized influence on the speech signal, such as emphasizing a particular word. For recognition tasks, this increases the challenge of isolating those feature characteristics modulated by emotion. Even humans are not perfect at perceiving the intended emotion for those emotional states that have similar acoustic characteristics. For instance, surprise can be perceived or understood as either joyous surprise (i.e., happiness) or apprehensive surprise (i.e., fear). Disgust is a form of disapproval and can be confused with anger.

There have been a few systems developed to synthesize emotional speech. The *Affect Editor* by Janet Cahn is among the earliest work in this area (Cahn, 1990). Her system was based on *DECtalk3,* a commercially available text-to-speech speech synthesizer. Given an English sentence and an emotional quality (one of anger, disgust, fear, joy, sorrow, or surprise), she developed a methodology for mapping the emotional correlates of speech (changes in pitch, timing, voice quality, and articulation) onto the underlying DECtalk synthesizer settings.

Table 11.1
Typical effect of emotions on adult human speech, adapted from Murray and Arnott (1993). The table has been extended to include some acoustic correlates of the emotion of surprise.

	Fear	Anger	Sorrow	Joy	Disgust	Surprise
Speech Rate	Much Faster	Slightly Faster	Slightly Slower	Faster or Slower	Very Much Slower	Much Faster
Pitch Average	Very Much Higher	Very Much Higher	Slightly Lower	Much Higher	Very Much Lower	Much Higher
Pitch Range	Much Wider	Much Wider	Slightly Narrower	Much Wider	Slightly Wider	
Intensity	Normal	Higher	Lower	Higher	Lower	Higher
Voice Quality	Irregular Voicing	Breathy Chest Tone	Resonant	Breathy Blaring	Grumbled Chest Tone	
Pitch Changes	Normal	Abrupt on Stressed Syllable	Downward Inflections	Smooth Upward Inflections	Wide Downward Terminal Inflections	Rising Contour
Articulation	Precise	Tense	Slurring	Normal	Normal	

She took great care to introduce the global prosodic effects of emotion while still preserving the more local influences of grammatical and lexical correlates of speech intonation. In a different approach Jun Sato (see `www.ee.seikei.ac.jp/user/junsato/research/`) trained a neural network to modulate a neutrally spoken speech signal (in Japanese) to convey one of four emotional states (happiness, anger, sorrow, disgust). The neural network was trained on speech spoken by Japanese actors. This approach has the advantage that the output speech signal sounds more natural than purely synthesized speech. It has the disadvantage, however, that the speech input to the system must be prerecorded.

With respect to giving Kismet the ability to generate emotive vocalizations, Cahn's work is a valuable resource. The DECtalk software gives us the flexibility to have Kismet generate its own utterance by assembling strings of phonemes (with pitch accents). I use Cahn's technique for mapping the emotional correlates of speech (as defined by her vocal affect parameters) to the underlying synthesizer settings. Because Kismet's vocalizations are at the proto-dialogue level, there is no grammatical structure. As a result, only producing the purely global emotional influence on the speech signal is noteworthy.

11.2   Expressive Voice Synthesis

Cahn's *vocal affect parameters (VAP)* alter the pitch, timing, voice quality, and articulation aspects of the speech signal. She documented how these parameter settings can be set to convey anger, fear, disgust, gladness, sadness, and surprise in synthetic speech. Emotions have a global impact on speech since they modulate the respiratory system, larynx, vocal tract, muscular system, heart rate, and blood pressure. The pitch-related parameters affect the pitch contour of the speech signal, which is the primary contributor for affective information. The pitch-related parameters include *accent shape, average pitch, pitch contour slope, final lowering, pitch range,* and *pitch reference line*. The timing-related parameters modify the prosody of the vocalization, often being reflected in speech rate and stress placement. The timing-related parameters include *speech rate, pauses, exaggeration,* and *stress frequency*. The voice-quality parameters include *loudness, brilliance, breathiness, laryngealization, pitch discontinuity,* and *pause discontinuity*. The articulation parameter modifies the precision of what is uttered, either being more enunciated or slurred. I describe these parameters in detail in the next section.

For Kismet, only some of these parameters are needed since several are inherently tied to sentence structure—the types and placement of pauses, for instance (see figure 11.1). In this section, I briefly describe those VAPs that are incorporated into Kismet's synthesized

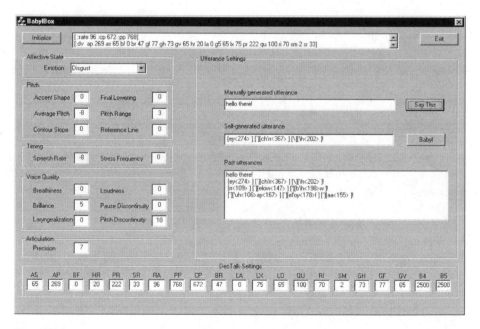

Figure 11.1
Kismet's expressive speech GUI. Listed is a selection of emotive qualities, the vocal affect parameters, and the synthesizer settings. A user can either manually enter an English phrase to be said, or can request automatically generated "Kismet-esque" babble. During run-time, Kismet operates in automatic generation mode.

Table 11.2
A description of the DECtalk synthesizer settings (see the DECtalk Software Reference Guide). Figure 11.3 illustrates the nominal pitch contour for neutral speech, and the net effect of changing these values for different expressive states. Cahn (1990) presents a detailed description of how each of these settings alters the pitch contour.

DECtalk Synthesizer Setting	Description
average pitch (Hz)	The average pitch of the pitch contour.
assertiveness (%)	The degree to which the voice tends to end statements with a conclusive fall.
baseline fall (Hz)	The desired fall (in Hz) of the baseline. The reference pitch contour around which all rule governed dynamic swings in pitch are about.
breathiness (dB)	Specifies the breathy quality of the voice due to the vibration of the vocal folds.
comma pause (ms)	Duration of pause due to a comma.
gain of frication	Gain of frication sound source.
gain of aspiration	Gain of aspiration sounds source.
gain of voicing	Gain of voicing sound source.
hat rise (Hz)	Nominal hat rise to the pitch contour plateau upon the first stressed syllable of the phrase. The hat-rise influence lasts throughout the phrase.
laryngealization (%)	Creaky voice. Results when the glottal pulse is narrow and the fundamental period is irregular.
loudness (dB)	Controls amplitude of speech waveform.
lax breathiness (%)	Specifies the amount of breathiness applied to the end of a sentence when going from voiced to voiceless sounds.
period pause (ms)	Duration of pause due to period.
pitch range (%)	Sets the range about the average pitch that the pitch contour expands and contracts. Specified in terms of percent of the nominal pitch range.
quickness (%)	Controls the speed of response to sudden requests to change pitch (due to pitch accents). Models the response time of the larynx.
speech rate (wpm)	Rate of speech in words per minute.
richness (%)	Controls the spectral change at lower frequencies (enhances the lower frequencies). Rich and brillant voices are more forceful.
smoothness (%)	Controls the amound of high frequency energy. There is less high frequency energy in a smooth voice. Varies inversely with brillance. Smoother voices sound friendlier.
stress rise (Hz)	The nominal height of the pitch rise and fall on each stressed syllable. This has a local influence on the contour about the stressed syllable.

speech. These vocal affect parameters modify the DECtalk synthesizer settings (summarized in table 11.2) according to the emotional quality to be expressed. The default values and max/min bounds for these settings are given in table 11.3. There is currently a single fixed mapping per emotional quality. Table 11.4 along with the equations presented in section 11.3 summarize how the vocal affect parameters are mapped to the DECtalk synthesizer settings. Table 11.5 summarizes how each emotional quality of voice is mapped onto the VAPs. Slight modifications in Cahn's specifications were made for Kismet—this should not be surprising as a different, more child-like voice was used. The discussion below motivates the mappings from VAPs to synthesizer settings as shown in figure 11.4. Cahn (1990) presents a detailed discussion of how these mappings were derived.

Table 11.3
Default DECtalk synthesizer settings for Kismet's voice (see the DECtalk Software Reference Guide). Section 11.3 describes the equations for altering these values to produce Kismet's expressive speech.

DECtalk Synthesizer Setting	Unit	Neutral Setting	Min Setting	Max Setting
average-pitch	Hz	306	260	350
assertiveness	%	65	0	100
baseline-fall	Hz	0	0	40
breathiness	dB	47	40	55
comma-pause	ms	160	−20	800
gain-of-frication	dB	72	60	80
gain-of-aspiration	dB	70	0	75
gain-of-voicing	dB	55	65	68
hat-rise	Hz	20	0	80
laryngealization	%	0	0	10
loudness	dB	65	60	70
lax-breathiness	%	75	100	0
period-pause	ms	640	−275	800
pitch-range	%	210	50	250
quickness	%	50	0	100
speech-rate	wpm	180	75	300
richness	%	40	0	100
smoothness	%	5	0	100
stress-rise	Hz	22	0	80

Pitch Parameters

The following six parameters influence the pitch contour of the spoken utterance. The pitch contour is the trajectory of the fundamental frequency, f_0, over time.

• *Accent Shape* Modifies the shape of the pitch contour for any pitch accented word by varying the rate of f_0 change about that word. A high accent shape corresponds to speaker agitation where there is a high peak f_0 and a steep rising and falling pitch contour slope. This parameter has a substantial contribution to DECtalk's stress-rise setting, which regulates the f_0 magnitude of pitch-accented words.

• *Average Pitch* Quantifies how high or low the speaker appears to be speaking relative to their normal speech. It is the average f_0 value of the pitch contour. It varies directly with DECtalk's average-pitch.

• *Contour Slope* Describes the general direction of the pitch contour, which can be characterized as rising, falling, or level. It contributes to two DECtalk settings. It has a small contribution to the assertiveness setting, and varies inversely with the baseline-fall setting.

• *Final Lowering* Refers to the amount that the pitch contour falls at the end of an utterance. In general, an utterance will sound emphatic with a strong final lowering, and tentative if

Table 11.4
Percent contributions of vocal affect parameters to DECtalk synthesizer settings. The absolute values of the contributions in the far right column add up to 1 (100%) for each synthesizer setting. See the equations in Section 11.3 for the mapping. The equations are similar to those used by Cahn (1990).

DECtalk Synthesizer Setting	DECtalk Symbol	Norm	Controlling Vocal Affect Parameter(s)	Percent of Control
average-pitch	ap	0.51	average pitch	1
assertiveness	as	0.65	final lowering	0.8
			contour direction	0.2
baseline-fall	bf	0	contour direction	−0.5
			final lowering	0.5
breathiness	br	0.46	breathiness	1
comma-pause	:cp	0.238	speech rate	−1
gain-of-frication	gf	0.6	precision of articulation	1
gain-of-aspiration	gh	0.933	precision of articulation	1
gain-of-voicing	gv	0.76	loudness	0.6
			precision of articulation	0.4
hat-rise	hr	0.2	reference line	1
laryngealization	la	0	laryngealization	1
loudness	lo	0.5	loudness	1
lax-breathiness	lx	0.75	breathiness	1
period-pause	:pp	0.67	speech rate	−1
pitch-range	pr	0.8	pitch range	1
quickness	qu	0.5	pitch discontinuity	1
speech-rate	:ra	0.2	speech rate	1
richness	ri	0.4	brillance	1
smoothness	sm	0.05	brillance	−1
stress-rise	sr	0.22	accent shape	0.8
			pitch discontinuity	0.2

weak. It can also be used as an auditory cue to regulate turn taking. A strong final lowering can signify the end of a speaking turn, whereas a speaker's intention to continue talking can be conveyed with a slight rise at the end. This parameter strongly contributes to DECtalk's `assertiveness` setting and somewhat to the `baseline-fall` setting.

• *Pitch Range* Measures the bandwidth between the maximum and minimum f_0 of the utterance. The pitch range expands and contracts about the average f_0 of the pitch contour. It varies directly with DECtalk's `pitch-range` setting.

• *Reference Line* Controls the reference pitch f_0 contour. Pitch accents cause the pitch trajectory to rise above or dip below this reference value. DECtalk's `hat-rise` setting very roughly approximates this.

Table 11.5
The mapping from each expressive quality of speech to the vocal affect parameters (VAPs). There is a single fixed mapping for each emotional quality.

Vocal Affect Parameter	Anger	Disgust	Fear	Happiness	Sorrow	Surprise	Neutral
accent shape	10	0	10	10	−7	9	0
average pitch	−10	−10	10	3	−7	6	0
contour slope	10	0	10	0	0	10	0
final lowering	10	5	−10	−4	8	−10	0
pitch range	10	5	10	10	−10	10	0
reference line	−10	0	10	−8	−1	−8	0
speech rate	4	−8	10	3	−6	6	0
stress frequency	0	0	10	5	1	0	0
breathiness	−5	0	0	−5	0	−9	0
brillance	10	5	10	−2	−6	9	0
laryngealization	0	0	−10	0	0	0	0
loudness	10	−5	10	8	−5	10	0
pause discontinuity	10	0	10	−10	−8	−10	0
pitch discontinuity	3	10	10	6	0	10	0
precision of articulation	10	7	0	−3	−5	0	0

Timing

The vocal affect timing parameters contribute to speech rhythm. Such correlates arise in emotional speech from physiological changes in respiration rate (changes in breathing patterns) and level of arousal.

• *Speech Rate* Controls the rate of words or syllables uttered per minute. It influences how quickly an individual word or syllable is uttered, the duration of sound to silence within an utterance, and the relative duration of phoneme classes. Speech is faster with higher arousal and slower with lower arousal. This parameter varies directly with DECtalk's `speech-rate` setting. It varies inversely with DECtalk's `period-pause` and `comma-pause` settings as faster speech is accompanied with shorter pauses.

• *Stress Frequency* Controls the frequency of occurrence of pitch accents and determines the smoothness or abruptness of f_0 transitions. As more words are stressed, the speech sounds more emphatic and the speaker more agitated. It filters other vocal affect parameters such as precision of articulation and accent shape, and thereby contributes to the associated DECtalk settings.

Voice Quality

Emotion can induce not only changes in pitch and tempo, but in voice quality as well. These phenomena primarily arise from changes in the larynx and articulatory tract.

• *Breathiness* Controls the aspiration noise in the speech signal. It adds a tentative and weak quality to the voice, when speaker is minimally excited. DECtalk `breathiness` and `lax-breathiness` vary directly with this.

• *Brillance* Controls the perceptual effect of relative energies of the high and low frequencies. When agitated, higher frequencies predominate and the voice is harsh or "brilliant". When speaker is relaxed or depressed, lower frequencies dominate and the voice sounds soothing and warm. DECtalk's `richness` setting varies directly as it enhances the lower frequencies. In contrast, DECtalk's `smoothness` setting varies inversely since it attenuates higher frequencies.

• *Laryngealization* Controls the perceived creaky voice phenomena. It arises from minimal sub-glottal pressure and a small open quotient such that f_0 is low, the glottal pulse is narrow, and the fundamental period is irregular. It varies directly with DECtalk's `laryngealization` setting.

• *Loudness* Controls the amplitude of the speech waveform. As a speaker becomes aroused, the sub-glottal pressure builds which increases the signal amplitude. As a result, the voice sounds louder. It varies directly with DECtalk's `loudness` setting. It also influences DECtalk's `gain-of-voicing`.

• *Pause Discontinuity* Controls the smoothness of f_0 transitions from sound to silence for unfilled pauses. Longer or more abrupt silences correlate with being more emotionally upset. It varies directly with DECtalk's `quickness` setting.

• *Pitch Discontinuity* Controls smoothness or abruptness of f_0 transitions, and the degree to which the intended targets are reached. With more speaker control, the transitions are smoother. With less control, they transitions are more abrupt. It contributes to DECtalk's `stress-rise` and `quickness` settings.

Articulation

The autonomic nervous system modulates articulation by inducing an assortment of physiological changes such as causing dryness of mouth or increased salivation.

• *Precision* Controls a range of articulation from enunciation to slurring. Slurring has minimal frication noise, whereas greater enunciation for consonants results in increased frication. Stronger enunciation also results in an increase in aspiration noise and voicing. The precision of articulation varies directly with DECtalk's `gain-of-frication`, `gain-of-voicing`, and `gain-of-aspiration`.

11.3   Implementation Overview

This section overviews the process for generating an expressive utterance and having Kismet say it with lip synchronization and facial animation. Figure 11.1 shows the controls of the expressive speech synthesizer as it appears on the NT machine. The active `emotion` is established by the emotion system as described in chapter 8. Currently, Kismet can vocalize an utterance with one of seven expressive states (anger, disgust, fear, happy, sad, surprise, and neutral). The decision to engage in vocal behavior is determined by the behavior system (chapter 9). This information is passed to the motor skills system where the request to speak with a given affective state is issued to the vocalization system. In the remainder of this chapter, I present how the vocalization system processes this request.

The algorithm for generating and performing an expressive Kismet-like vocalization is as follows:

1. Update vocal affect parameters based on current `emotion`.
2. Map from vocal affect parameters to synthesizer settings.
3. Generate the utterance to speak.
4. Assemble the full command and send it to the synthesizer.
5. Extract features from speech signal for lip synchronization.
6. Send the speech signal to the sound card.
7. Execute lip synchronization movements.

Mapping Vocal Affect Parameters to Synthesizer Settings

The vocal affect parameters outlined in section 11.2 are derived from the acoustic correlates of emotion in human speech. To have DECtalk produce these effects in synthesized speech, these vocal affect parameters must be computationally mapped to the underlying synthesizer settings. There is a single fixed mapping per emotional quality. With some minor modifications, Cahn's mapping functions are adapted to Kismet's implementation.

The vocal affect parameters can assume integer values within the range of $(-10, 10)$. Negative numbers correspond to lesser effects, positive numbers correspond to greater effects, and zero is the neutral setting. These values are set according to the current specified emotion as shown in table 11.5.

Linear changes in these parameter values result in a non-linear change in synthesizer settings. Furthermore, the mapping between parameters and synthesizer settings is not necessarily one-to-one. Each parameter affects a percent of the final synthesizer setting's value (table 11.4). When a synthesizer setting is modulated by more than one parameter, its final value is the sum of the effects of the controlling parameters. The total of the absolute values of these percentages must be 100%. See table 11.3 for the allowable bounds of synthesizer settings. The computational mapping occurs in three stages.

In the first stage, the percentage of each of the vocal affect parameters (VAP_i) to its total range is computed, (PP_i). This is given by the equation:

$$PP_i = \frac{VAP_{value_i} + VAP_{offset}}{VAP_{max} - VAP_{min}}$$

VAP_i is the current VAP under consideration, VAP_{value} is its value specified by the current emotion, $VAP_{offset} = 10$ adjusts these values to be positive, $VAP_{max} = 10$, and $VAP_{min} = -10$.

In the second stage, a weighted contribution ($WC_{j,i}$) of those VAP_i that control each of DECtalk's synthesizer settings (SS_j) is computed. The far right column of table 11.4 specifies each of the corresponding *scale factors* ($SF_{j,i}$). Each scale factor represents a percentage of control that each VAP_i applies to its synthesizer setting SS_j.

For each synthesizer setting, SS_j:
    For each corresponding scale factor, $SF_{j,i}$ of VAP_i:
        If $SF_{j,i} > 0$
            $WC_{j,i} = PP_i \times SF_{j,i}$
        If $SF_{j,i} \leq 0$
            $WC_{j,i} = (1 - PP_i) \times (-SF_{j,i})$
    $SS_j = \sum_i WC_{j,i}$

At this point, each synthesizer value has a value $0 \leq SS_j \leq 1$. In the final stage, each synthesizer setting SS_j is scaled about 0.5. This produces the final synthesizer value, $SS_{j_{final}}$. The final value is sent to the speech synthesizer. The maximum, minimum, and default values of the synthesizer settings are shown in table 11.3.

For each final synthesizer setting, $SS_{j_{final}}$:
    Compute $SS_{j_{offset}} = SS_j - norm$
    If $SS_{j_{offset}} > 0$
        $SS_{j_{final}} = SS_{j_{default}} + (2 \times SS_{j_{offset}} \times (SS_{j_{max}} - SS_{j_{min}}))$
    If $SS_{j_{offset}} \leq 0$
        $SS_{j_{final}} = SS_{j_{default}} + (2 \times SS_{j_{offset}} \times (SS_{j_{default}} - SS_{j_{min}}))$

Generating the Utterance

To engage in proto-dialogues with its human caregiver and to partake in vocal play, Kismet must be able to generate its own utterances. The algorithm outlined below produces a style of speech that is reminiscent of a tonal dialect. As it stands, the output is quite distinctive and contributes significantly to Kismet's personality (as it pertains to its manner of vocal expression). It is really intended, however, as a placeholder for a more sophisticated utterance generation algorithm to eventually replace it. In time, Kismet will be able to adjust its utterance based on what it hears, but this is the subject of future work.

Based upon DECtalk's phonemic speech mode, the generated string to be synthesized is assembled from pitch accents, phonemes, and end syntax. The end syntax is a requirement of DECtalk and does not serve a grammatical function. However, as with the pitch accents, it does influence the prosody of the utterance and is used in this manner. The DECtalk phonemes are summarized in table 11.6 and the accents are summarized in table 11.7.

Table 11.6
DECtalk phonemes for generating utterances.

Consonants				Vowels		Vowels	
b	**b**et	n	**n**et	aa	b**o**b	oy	b**oy**
ch	**ch**in	nx	si**ng**	ae	b**a**t	rr	b**ir**d
d	**d**ebt	p	**p**et	ah	b**u**t	uh	b**oo**k
dh	**th**is	r	**r**ed	ao	b**ou**ght	uw	l**u**te
el	bott**le**	s	**s**it	aw	b**ou**t	yu	c**u**te
en	butt**on**	sh	**sh**in	ax	**a**bout		allophones
f	**f**in	t	**t**est	ay	b**i**te	dx	ri**d**er
g	**g**uess	th	**th**in	eh	b**e**t	lx	wi**ll**
hx	**h**ead	v	**v**est	ey	b**a**ke	q	we **ea**t
jh	**g**in	w	**w**et	ih	b**i**t	rx	o**r**ation
k	**k**en	yx	**y**et	ix	kiss**es**	tx	La**t**in
l	**l**et	z	**z**oo	iy	b**ea**t		Silence
m	**m**et	zh	a**z**ure	ow	b**oa**t		_(underscore)

Table 11.7
DECtalk accents and end syntax for generating utterances.

Symbol	Name	Indicates	Symbol	Name	Indicates
[']	apostrophe	primary stress	[,]	comma	clause boundaries
[']	grave accent	secondary stress	[.]	period	period
["]	quotation mark	emphatic stress	[?]	question mark	question mark
[/]	slash	pitch rise	[!]	exclamation mark	exclamation mark
[\]	backslash	pitch fall	[ ]	space	word boundary
[/ \]	hat	pitch rise and fall			

Kismet's vocalizations are generated as follows:

Randomly choose number of proto-words, $getUtteranceLength() = length_{utterance}$
For $i = (0, length_{utterance})$, generate a proto-word, $protoWord$
    Generate a ($wordAccent$, $word$) pair
    Randomly choose word accent, $getAccent()$
    Randomly choose number of syllables of proto-word, $getWordLength() = length_{word}$
    Choose which syllable receives primary stress, $assignStress()$
    For $j = (0, length_{word})$, generate a syllable
        Randomly choose the type of syllable, $syllableType$
        if $syllableType = vowelOnly$
          if this syllable has primary stress
            then $syllable = getStress() + getVowel() + getDuration()$
          else $syllable = getVowel() + getDuration()$
        if $syllableType = consonantVowel$
          if this syllable has primary stress
            then $syllable = getConsonant() + getStress() + getVowel() +$
                    $getDuration()$
          else $syllable = getConsonant() + getVowel() + getDuration()$
        if $syllableType = consonantVowelConsonant$
          if this syllable has primary stress
            then $syllable = getConsonant() + getStress() + getVowel() +$
                    $getDuration() + getConsonant()$
          else $syllable = getConsonant() + getVowel() + getDuration() +$
                    $getConsonant()$
        if $syllableType = vowelVowel$
          if this syllable has primary stress
            then $syllable = getStress() + getVowel() + getDuration() + getvowel() +$
                    $getDuration()$
          else $syllable = getVowel() + getDuration() + getVowel() +$
                    $getDuration()$
        $protoWord = append(protoWord, syllable)$
      $protoWord = append(wordAccent, protoWord)$
$utterance = append(utterance, protoWord)$

  Where:

• $GetUtteranceLength()$ randomly chooses a number between $(1, 5)$. This specifies the number of proto-words in a given utterance.

• *GetWordLength*() randomly chooses a number between (1, 3). This specifies the number of syllables in a given proto-word.

• *GetPunctuation*() randomly chooses one of end syntax markers as shown in table 11.7. This is biased by emotional state to influence the end of the pitch contour.

• *GetAccent*() randomly choose one of six accents (including no accent) as shown in table 11.7.

• *assignStress*() selects which syllable receives primary stress.

• *getVowel*() randomly choose one of eighteen vowel phonemes as shown in table 11.6.

• *getConsonant*() randomly chooses one of twenty-six consonant phonemes as shown in table 11.6.

• *getStress*() gets the primary stress accent.

• *getDuration*() randomly chooses a number between (100, 500) that specifies the vowel duration in msec. This selection is biased by the emotional state where lower arousal vowels tend to have longer duration, and high arousal states have shorter duration.

11.4   Kismet's Expressive Utterances

Given the phonemic string to be spoken and the updated synthesizer settings, Kismet can vocally express itself with different emotional qualities. To evaluate Kismet's speech, the produced utterances are analyzed with respect to the acoustical correlates of emotion. This will reveal if the implementation produces similar acoustical changes to the speech waveform given a specified emotional state. It is also important to evaluate how the affective modulations of the synthesized speech are perceived by human listeners.

Analysis of Speech

To analyze the performance of the expressive vocalization system, the dominant acoustic features that are highly correlated with emotive state were extracted. The acoustic features and their modulation with `emotion` are summarized in table 11.1. Specifically, these are average pitch, pitch range, pitch variance, and mean energy. To measure speech rate, the overall time to speak and the total time of voiced segments were determined.

These features were extracted from three phrases:

• *Look at that picture*

• *Go to the city*

• *It's been moved already*

Table 11.8
Table of acoustic features for the three utterances.

	nzpmean	nzpvar	pmax	pmin	prange	egmean	length	voiced	unvoiced
anger-city	292.5	6348.7	444.4	166.7	277.7	112.2	81	52	29
anger-moved	269.1	4703.8	444.4	160	284.4	109.8	121	91	30
anger-picture	273.2	6850.3	444.4	153.8	290.6	110.2	112	51	61
anger-average	278.3	5967.6	444.4	160.17	284.2	110.7	104.6	64.6	40
calm-city	316.8	802.9	363.6	250	113.6	102.6	85	58	27
calm-moved	304.5	897.3	363.6	266.7	96.9	103.6	124	94	30
calm-picture	302.2	1395.5	363.6	235.3	128.3	102.4	118	73	45
calm-average	307.9	1031.9	363.6	250.67	112.93	102.9	109	75	34
disgust-city	268.4	2220.0	400	173.9	226.1	102.5	124	83	41
disgust-moved	264.6	1669.2	400	190.5	209.5	101.6	173	123	50
disgust-picture	275.2	3264.1	400	137.9	262.1	102.3	157	82	75
disgust-average	269.4	2384.4	400	167.4	232.5	102.1	151.3	96	55.3
fear-city	417.0	8986.7	500	235.3	264.7	102.8	59	27	32
fear-moved	357.2	7145.5	500	160	340	102.6	89	53	36
fear-picture	388.2	8830.9	500	160	340	103.6	86	41	45
fear-average	387.4	8321.0	500	185.1	314.9	103.0	78	40.3	37.6
happy-city	388.3	5810.6	500	285.7	214.3	106.6	71	54	17
happy-moved	348.2	6188.8	500	173.9	326.1	109.2	109	78	31
happy-picture	357.7	6038.3	500	266.7	233.3	106.0	100	57	43
happy-average	364.7	6012.6	500	242.1	257.9	107.2	93.3	63	30.3
sad-city	279.8	77.9	285.7	266.7	19	98.6	88	62	26
sad-moved	276.9	90.7	285.7	266.7	19	99.1	144	93	51
sad-picture	275.5	127.2	285.7	250	35.7	98.3	138	83	55
sad-average	277.4	96.6	285.7	261.1	24.5	98.7	123.3	79.3	44
surprise-city	394.3	8219.4	500	148.1	351.9	107.5	69	49	20
surprise-moved	360.3	7156.0	500	160	340	107.8	101	84	17
surprise-picture	371.6	8355.7	500	285.7	214.3	106.7	98	54	44
surprise-average	375.4	7910.4	500	197.9	302.0	107.3	89.3	62.3	27

The results are summarized in table 11.8. The values for each feature are displayed for each phrase with each emotive quality (including the neutral state). The averages are also presented in the table and plotted in figure 11.2. These plots easily illustrate the relationship of how each emotive quality modulates these acoustic features with respect to one another. The pitch contours for each emotive quality are shown in figure 11.3. They correspond to the utterance "It's been moved already."

Relating these plots with table 11.1, it is clear that many of the acoustic correlates of emotive speech are preserved in Kismet's speech. I have made several incremental adjustments to the qualities of Kismet's speech according to what was learned from subject evaluations. The final implementation differs in some cases from table 11.1 (as noted below), but the results show a dramatic improvement in subject recognition performance from earlier evaluations.

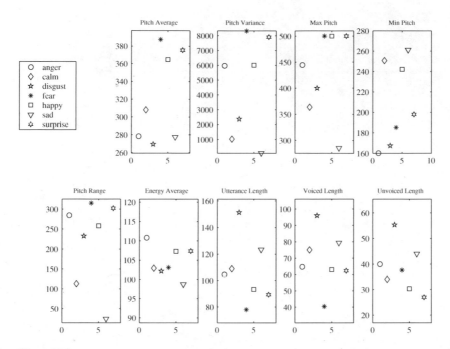

Figure 11.2
Plots of acoustic features of Kismet's speech. Plots illustrate how each emotion relates to the others for each acoustic feature. The horizontal axis simply maps an integer value to each emotion for ease of viewing (anger = 1, calm = 2, etc.)

Kismet's vocal quality varies with its "emotive" state as follows:

• Fearful speech is very fast with wide pitch contour, large pitch variance, very high mean pitch, and normal intensity. I have added a slightly breathy quality to the voice as people seem to associate it with a sense of trepidation.

• Angry speech is loud and slightly fast with a wide pitch range and high variance. I've purposefully implemented a low mean pitch to give the voice a prohibiting quality. This differs from table 11.1, but a preliminary study demonstrated a dramatic improvement in recognition performance of naive subjects. This makes sense as it gives the voice a threatening quality.

• Sad speech has a slower speech rate, with longer pauses than normal. It has a low mean pitch, a narrow pitch range and low variance. It is softly spoken with a slight breathy quality. This differs from table 11.1, but it gives the voice a tired quality. It has a pitch contour that falls at the end.

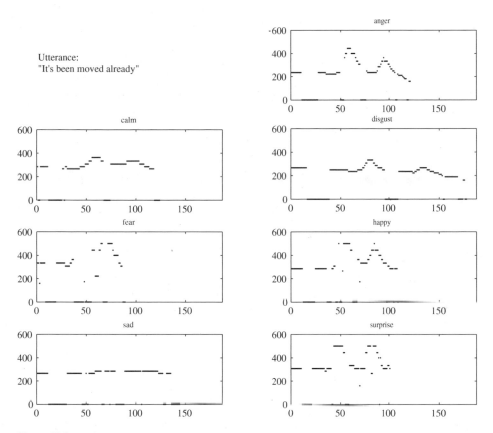

Figure 11.3
Pitch analysis of Kismet's speech for the English phrase "It's been moved already."

• Happy speech is relatively fast, with a high mean pitch, wide pitch range, and wide pitch variance. It is loud with smooth undulating inflections as shown in figure 11.3.

• Disgusted speech is slow with long pauses interspersed. It has a low mean pitch with a slightly wide pitch range. It is fairly quiet with a slight creaky quality to the voice. The contour has a global downward slope as shown in figure 11.3.

• Surprised speech is fast with a high mean pitch and wide pitch range. It is fairly loud with a steep rising contour on the stressed syllable of the final word.

Human Listener Experiments

To evaluate Kismet's expressive speech, nine subjects were asked to listen to prerecorded utterances and to fill out a forced-choice questionnaire. Subjects ranged from 23 to 54 years of age, all affiliated with MIT. The subjects had very limited to no familiarity with Kismet's voice.

In this study, each subject first listened to an introduction spoken with Kismet's neutral expression. This was to acquaint the subject with Kismet's synthesized quality of voice and neutral affect. A series of eighteen utterances followed, covering six expressive qualities (anger, fear, disgust, happiness, surprise, and sorrow). Within the experiment, the emotive qualities were distributed randomly. Given the small number of subjects per study, I only used a single presentation order per experiment. Each subject could work at his/her own pace and control the number of presentations of each stimulus.

The three stimulus phrases were: "I'm going to the city," "I saw your name in the paper," and "It's happening tomorrow." The first two test phrases were selected because Cahn had found the word choice to have reasonably neutral affect. In a previous version of the study, subjects reported that it was just as easy to map emotional correlates onto English phrases as to Kismet's randomly generated babbles. Their performance for English phrases and Kismet's babbles supports this. We believed it would be easier to analyze the data to discover ways to improve Kismet's performance if a small set of fixed English phrases were used.

The subjects were simply asked to circle the word which best described the voice quality. The choices were "anger," "disgust," "fear/panic," "happy," "sad," "surprise/excited." From a previous iteration of the study, I found that word choice mattered. A given emotion category can have a wide range of vocal affects. For instance, the subject could interpret "fear" to imply "apprehensive," which might be associated with Kismet's whispery vocal expression for sadness. Alternatively, it could be associated with "panic" which is a more aroused interpretation. The results from these evaluations are summarized in table 11.9.

Overall, the subjects exhibited reasonable performance in correctly mapping Kismet's expressive quality with the targeted emotion. However, the expression of "fear" proved

Table 11.9
Naive subjects assessed the emotion conveyed in Kismet's voice in a forced-choice evaluation. The emotional qualities were recognized with reasonable performance except for "fear" which was most often confused for "surprise/excitement." Both expressive qualities share high arousal, so the confusion is not unexpected.

	anger	disgust	fear	happy	sad	surprise	% correct
anger	75	15	0	0	0	10	75
disgust	21	50	4	0	25	0	50
fear	4	0	25	8	0	63	25
happy	0	4	4	67	8	17	67
sad	8	8	0	0	84	0	84
surprise	4	0	25	8	4	59	59

Forced-Choice Percentage (random = 17%)

problematic. For all other expressive qualities, the performance was significantly above random. Furthermore, misclassifications were highly correlated to similar emotions. For instance, "anger" was sometimes confused with "disgust" (sharing negative valence) or "surprise/excitement" (both sharing high arousal). "Disgust" was confused with other negative emotions. "Fear" was confused with other high arousal emotions (with "surprise/excitement" in particular). The distribution for "happy" was more spread out, but it was most often confused with "surprise/excitement," with which it shares high arousal. Kismet's "sad" speech was confused with other negative emotions. The distribution for "surprise/excitement" was broad, but it was most often confused for "fear."

Since this study, the vocal affect parameter values have been adjusted to improve the distinction between "fear" and "surprise." Kismet's fearful affect has gained a more apprehensive quality by lowering the volume and giving the voice a slightly raspy quality (this was the version that was analyzed in section 11.4). In a previous study I found that people often associated the raspy vocal quality with whispering and apprehension. "Surprise" has also been enhanced by increasing the amount of stress rise on the stressed syllable of the final word. Cahn analyzed the sentence structure to introduce irregular pauses into her implementation of "fear." This makes a significant contribution to the interpretation of this emotional state. In practice, however, Kismet only babbles, so modifying the pausing via analysis of sentence structure is premature as sentences do not exist.

Given the number and homogeneity of subjects, I cannot make strong claims regarding Kismet's ability to convey emotion through expressive speech. More extensive studies need to be carried out, yet, for the purposes of evaluation, the current set of data is promising. Misclassifications are particularly informative. The mistakes are highly correlated with similar emotions, which suggests that arousal and valence are conveyed to people (arousal being more consistently conveyed than valence). I am using the results of this study to improve Kismet's expressive qualities. In addition, Kismet expresses itself through multiple modalities, not just through voice. Kismet's facial expression and body posture should help resolve the ambiguities encountered through voice alone.

11.5   Real-Time Lip Synchronization and Facial Animation

Given Kismet's ability to express itself vocally, it is important that the robot also be able to support this vocal channel with coordinated facial animation. This includes synchronized lip movements to accompany speech along with facial animation to lend additional emphasis to the stressed syllables. These complementary motor modalities greatly enhance the robot's delivery when it speaks, giving the impression that the robot "means" what it says. This makes the interaction more engaging for the human and facilitates proto-dialogue.

Guidelines from Animation

The earliest examples of lip synchronization for animated characters dates back to the 1940's in classical animation (Blair, 1949), and back to the 1970s for computer-animated characters (Parke, 1972). In these early works, all of the lip animation was crafted by hand (a very time-consuming process). Over time, a set of guidelines evolved that are largely adhered to by animation artists today (Madsen, 1969).

According to Madsen, *simplicity is the secret to successful lip animation*. Extreme accuracy for cartoon animation often looks forced or unnatural. Thus, the goal in animation is not to always imitate realistic lip motions, but *to create a visual shorthand that passes unchallenged by the viewer* (Madsen, 1969). As the realism of the character increases, however, the accuracy of the lip synchronization follows.

Kismet is a fanciful and cartoon-like character, so the guidelines for cartoon animation apply. In this case, the guidelines suggest that the animator focus on vowel lip motions (especially o and w) accented with consonant postures (m, b, p) for lip closing. Precision of these consonants gives credibility to the generalized patterns of vowels. The transitions between vowels and consonants should be reasonable approximations of lip and jaw movement. Fortunately, more latitude is granted for more fanciful characters. The mechanical response time of Kismet's lip and jaw motors places strict constraints on how fast the lips and jaw can transition from posture to posture. Madsen also stresses that care must be taken in conveying emotion, as the expression of voice and face can change dramatically.

Extracting Lip Synch Info

To implement lip synchronization on Kismet, a variety of information must be computed in real-time from the speech signal. By placing DECtalk in *memory mode* and issuing the command string (utterance with synthesizer settings), the DECtalk software generates the speech waveform and writes it to memory (a 11.025 kHz waveform). In addition, DECtalk extracts time-stamped phoneme information. From the speech waveform, one can compute its time-varying energy over a window size of 335 samples, taking care to synchronize the phoneme and energy information, and send ($phoneme[t]$, $energy[t]$) pairs to the QNX machine at 33 Hz to coordinate jaw and lip motor control. A similar technique using DECtalk's phoneme extraction capability is reported by Waters and Levergood (1993) for real-time lip synchronization for computer-generated facial animation.

To control the jaw, the QNX machine receives the phoneme and energy information and updates the commanded jaw position at 10 Hz. The mapping from energy to jaw opening is linear, bounded within a range where the minimum position corresponds to a closed mouth, and the maximum position corresponds to an open mouth characteristic of surprise. Using only energy to control jaw position produces a lively effect but has its limitations (Parke & Waters, 1996). For Kismet, the phoneme information is used to make sure that the jaw is closed when either a *m*, *p*, or *b* is spoken or there is silence. This may not necessarily be the case if only energy were used.

Upon receiving the phoneme and energy information from the vocalization system, the QNX vocal communication process passes this information to the motor skill system via the DPRAM. The motor skill system converts the energy information into a measure of facial emphasis (linearly scaling the energy), which is then passed onto the lip synchronization and facial animation processes of the face control motor system. The motor skill system also maps the phoneme information onto lip postures and passes this information to the lip synchronization and facial animation processes of the motor system that controls the face (described in chapter 10). Figure 11.4 illustrates the stages of computation from the raw speech signal to lip posture, jaw opening, and facial emphasis.

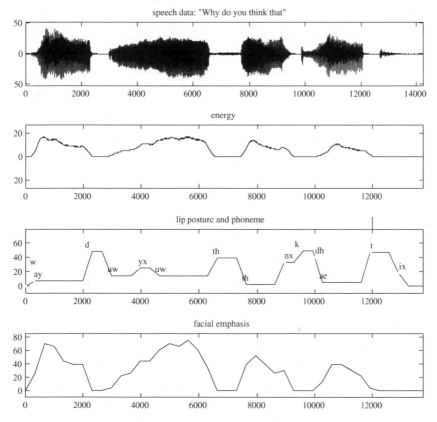

Figure 11.4
Plot of speech signal, energy, phonemes/lip posture, and facial emphasis for the phrase "Why do you think that?" Time is in 0.1 ms increments. The total amount of time to vocalize the phrase is 1.4 sec.

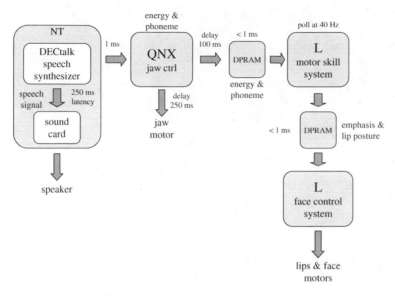

Figure 11.5
Schematic of the flow of information for lip synchronization. This figure illustrates the latencies of the system and the compensatory delays to maintain synchrony.

The computer network involved in lip synchronization is a bit convoluted, but supports real-time performance. Figure 11.5 illustrates the information flow through the system and denotes latencies. Within the NT machine, there is a latency of approximately 250 ms from the time the synthesizer generates the speech signal and extracts phoneme information until that speech signal is sent to the sound card. Immediately following the generation and feature extraction phase, the NT machine sends this information to the QNX node that controls the jaw motor. The latency of this stage is less than 1 ms. Within QNX, the energy signal and phoneme information are used to compute the jaw position. To synchronize jaw movement with sound production from the sound card, the jaw command position is delayed by 250 ms. For the same reason, the QNX machine delays the transfer of energy and phoneme information by 100 ms to the L-based machines. Dual-ported RAM communication is sub-millisecond. The lip synchronization processes running on L polls and updates their energy and phoneme values at 40 Hz, much faster than the phoneme information is changing and much faster than the actuators can respond. Energy is scaled to control the amount of facial emphasis, and the phonemes are mapped to lip postures. The lip synchronization performance is well-coordinated with speech output since the delays and latencies are fairly consistent.

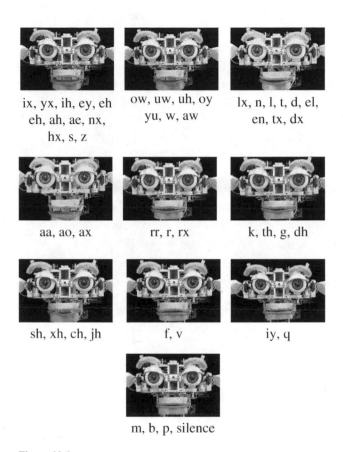

Figure 11.6
Kismet's mapping of lip postures to phonemes.

Kismet's ability to lip-sync within its limits greatly enhances the perception that it is genuinely talking (instead of being some disembodied speech system). It also contributes to the life-like quality and charm of the robot's behavior.

Figure 11.6 shows how the fifty DECtalk phonemes are mapped to Kismet's lip postures. Kismet obviously has a limited repertoire as it cannot make many of the lip movements that humans do. For instance, it cannot protrude its lips (important for *sh* and *ch* sounds), nor does it have a tongue (important for *th* sounds), nor teeth. However, computer-animated lip synchronization often maps the 45 distinct English phonemes onto a much more restricted set of visually distinguishable lip postures; eighteen is preferred (Parke & Waters, 1996). For cartoon characters, a subset of ten lip and jaw postures is enough for reasonable artistic conveyance (Fleming & Dobbs, 1999). Kismet's ten lip postures tend toward the absolute minimal set specified by Fleming and Dobbs (1999), but is reasonable given its physical appearance. As the robot speaks, new lip posture targets are specified at 33 Hz. Since the phonemes do not change this quickly, many of the phonemes repeat. There is an inherent limit in how fast Kismet's lip and jaw motors can move to the next commanded, so the challenge of co-articulation is somewhat addressed of by the physics of the motors and mechanism.

Lip synchronization is only part of the equation, however. Faces are not completely still when speaking, but move in synchrony to provide emphasis along with the speech. Using the energy of the speech signal to animate Kismet's face (along with the lips and jaw) greatly enhances the impression that Kismet "means" what it says. For Kismet, the energy of the speech signal influences the movement of its eyelids and ears. Larger speech amplitudes result in a proportional widening of the eyes and downward pulse of the ears. This adds a nice degree of facial emphasis to accompany the stress of the vocalization.

Since the speech signal influences facial animation, the emotional correlates of facial posture must be blended with the animation arising from speech. How this is accomplished within the face control motor system is described at length in chapter 10. The emotional expression establishes the baseline facial posture about which all facial animation moves. The current "emotional" state also influences the speed with which the facial actuators move (lower arousal results in slower movements, higher arousal results in quicker movements). In addition, `emotions` that correspond to higher arousal produce more energetic speech, resulting in bigger amplitude swings about the expression baseline. Similarly, `emotions` that correspond to lower arousal produce less energetic speech, which results in smaller amplitudes. The end product is a highly expressive and coordinated movement of face with voice. For instance, angry sounding speech is accompanied by large and quick twitchy movements of the ears eyelids. This undeniably conveys agitation and irritation. In contrast, sad sounding speech is accompanied by slow, droopy, listless movements of the ears and eyelids. This conveys a forlorn quality that often evokes sympathy from the human observer.

11.6   Limitations and Extensions

Kismet's expressive speech can certainly be improved. In the current implementation I have only included those acoustic correlates that have a global influence on the speech signal and do not require local analysis of the sentence structure. I currently modulate voice quality, speech rate, pitch range, average pitch, intensity, and the global pitch contour. Data from naive subjects is promising, although more could certainly be done. I have done very little with changes in articulation. The precision or imprecision of articulation could be enhanced by substituting voiced for unvoiced phonemes as Cahn describes in her thesis.

By analyzing sentence structure, several more influences can be introduced. For instance, carefully selecting the types of stress placed on emphasized and de-emphasized words, as well as introducing different kinds of pausing, can be used to strengthen the perception of negative `emotions` such as `fear`, `sadness`, and `disgust`. Given the immediate goal of proto-language, there is no sentence structure to analyze. Nonetheless, to extend Kismet's expressive abilities to English sentences, the grammatical and lexical constraints must be carefully considered.

On a slightly different vein, emotive sounds such as laughter, cries, coos, gurgles, screams, shrieks, yawns, and so forth could be introduced. DECtalk supports the ability to play pre-recorded sound files. An initial set of emotive sounds could be modulated to add variability.

Extensions to Utterance Generation

Kismet's current manner of speech has wide appeal to those who have interacted with the robot. There is sufficient variability in phoneme, accent, and end syntax choice to permit an engaging proto-dialogue. If Kismet's utterance has the intonation of a question, people will treat it as such—often "re-stating" the question as an English sentence and then answering it. If Kismet's intonation has the intonation of a statement, they respond accordingly. They may say something such as, "Oh, I see," or perhaps issue another query such as, "So then what did you do?" The utterances are complex enough to sound as if the robot is speaking a different language.

Even so, the current utterance generation algorithm is really intended as a placeholder for a more sophisticated generation algorithm. There is interest in computationally modeling canonical babbling so that the robot makes vocalizations characteristic of an eight-month old child (de Boysson-Bardies, 1999). This would significantly limit the range of the utterances the robot currently produces, but would facilitate the acquisition of proto-language. Kismet varies many parameters at once, so the learning space is quite large. By modeling canonical babbling, the robot can systematically explore how a limited set of parameters modulates the way its voice sounds. Introducing variations upon a theme during vocal games with the caregiver as well as on its own could simplify the learning process (see chapters 2 and 3). By interfacing what the robot vocally generates with what it hears, the robot could begin to explore its vocal capabilities, how to produce targeted effects, and how these utterances influence the caregiver's behavior.

Improvements to Lip Synchronization

Kismet's lip synchronization and facial animation are compelling and well-matched to Kismet's behavior and appearance. The current implementation, however, could be improved upon and extended in a couple of ways. First, the latencies throughout the system could be reduced. This would give us tighter synchronization. Higher performance actuators could be incorporated to allow a faster response time. This would also support more precise lip synchronization.

A tongue, teeth, and lips that could move more like those of a human would add more realism. This degree of realism is unnecessary for purposes here, however, and is tremendously difficult to achieve. As it stands, Kismet's lip synchronization is a successful shorthand that goes unchallenged by the viewer.

11.7   Summary

Kismet uses an expressive vocalization system that can generate a wide range of utterances. This system addresses issues regarding the expressiveness and richness of Kismet's vocal modality, and how it supports social interaction. I have found that the vocal utterances are rich enough to facilitate interesting proto-dialogues with people, and that the expressiveness of the voice is reasonably identifiable. Furthermore, the robot's speech is complemented by real-time animated facial animation that enhances delivery. Instead of trying to achieve realism, this system is well-matched with the robot's whimsical appearance and limited capabilities. The end result is a well-orchestrated and compelling synthesis of voice, facial animation, and affect that make a significant contribution to the expressiveness and personality of the robot.

12 Social Constraints on Animate Vision

The control of animate vision for a social robot poses challenges beyond issues of stability and accuracy, as well as advantages beyond computational efficiency and perceptual robustness (Ballard, 1989). Kismet's human-like eye movements have high communicative value to the people that interact with it. Hence the challenge of interacting with humans constrains how Kismet appears physically, how it moves, how it perceives the world, and how its behaviors are organized. This chapter describes Kismet's integrated *visual-motor system*. The system must negotiate between the physical constraints of the robot, the perceptual needs of the robot's behavioral and motivational systems, and the social implications of motor acts. It presents those systems responsible for generating Kismet's compelling visual behavior.

From a social perspective, human eye movements have a high communicative value (as illustrated in figure 12.1). For example, gaze direction is a good indicator of the locus of visual attention. I have discussed this at length in chapter 6. The dynamic aspects of eye movement, such as staring versus glancing, also convey information. Eye movements are particularly potent during social interactions, such as conversational turn-taking, where making and breaking eye contact plays an important role in regulating the exchange. We model the eye movements of our robots after humans, so that they may have similar communicative value.

From a functional perspective, the human system is so good at providing a stable percept of the world that we have no intuitive appreciation of the physical constraints under which it operates. Fortunately, there is a wealth of data and proposed models for how the human visual system is organized (Kandel et al., 2000). This data provides not only a modular decomposition but also mechanisms for evaluating the performance of the complete system.

12.1 Human Visual Behavior

Kismet's visual-motor control is modeled after the human oculo-motor system. By doing so, my colleagues and I hope to harness both the computational efficiency and perceptual robustness advantages of an animate vision system, as well as the communicative power of human eye movements. In this section I briefly survey the key aspects of the human visual system used as a guideline to design Kismet's visual apparatus and eye movement primitives.

Foveate vision Humans have *foveate* vision. The fovea (the center of the retina) has a much higher density of photoreceptors than the periphery. This means that to see an object clearly, humans must move their eyes such that the image of the object falls on the fovea. The advantage of this receptor layout is that humans enjoy both a wide peripheral field of view as well as high acuity vision. The wide field of view is useful for directing visual attention to interesting features in the environment that may warrant further detailed analysis. This

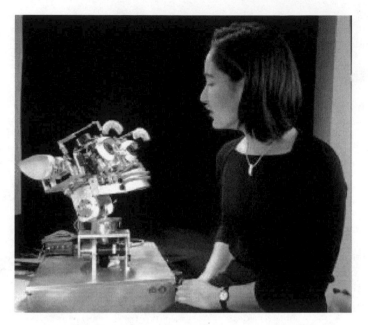

Figure 12.1
Kismet is capable of conveying intentionality through facial expressions and behavior. Here, the robot's physical state expresses attention to and interest in the human beside it. Another person—for example, the photographer—would expect to have to attract the robot's attention before being able to influence its behavior.

analysis is performed while directing gaze to that target and using foveal vision for detailed processing over a localized region of the visual field.

Vergence movements   Humans have binocular vision. The visual disparity of the images from each eye give humans one visual cue to perceive depth (humans actually use multiple cues (Kandel et al., 2000)). The eyes normally move in lock-step, making equal, conjunctive movements. For a close object, however, the eyes need to turn towards each other somewhat to correctly image the object on the foveae of the two eyes. These disjunctive movements are called *vergence* and rely on depth perception (see figure 12.2).

Saccades   Human eye movement is not smooth. It is composed of many quick jumps, called *saccades,* which rapidly re-orient the eye to project a different part of the visual scene onto the fovea. After a saccade, there is typically a period of fixation, during which the eyes are relatively stable. They are by no means stationary, and continue to engage in corrective micro-saccades and other small movements. Periods of fixation typically end after some hundreds of milliseconds, after which a new saccade will occur.

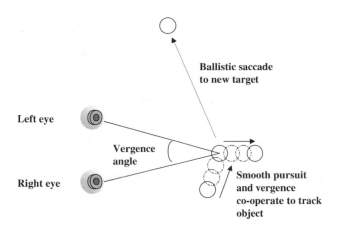

Figure 12.2
The four characteristic types of human eye motion.

Smooth pursuit   If, however, the eyes fixate on a moving object, they can follow it with a continuous tracking movement called *smooth pursuit*. This type of eye movement cannot be evoked voluntarily, but only occurs in the presence of a moving object.

Vestibulo-ocular reflex and opto-kinetic response   Since eyes also move with respect to the head, they need to compensate for any head movements that occur during fixation. The *vestibulo-ocular reflex* (VOR) uses inertial feedback from the vestibular system to keep the orientation of the eyes stable as the eyes move. This is a very fast response, but is prone to the accumulation of error over time. The *opto-kinetic nystagmus* (OKN) is a slower compensation mechanism that uses a measure of the visual slip of the image across the retina to correct for drift. These two mechanisms work together to give humans stable gaze as the head moves.

12.2   Design Issues for Visual Behavior

Kismet is endowed with visual perception and visual motor abilities that are human-like in their physical implementation. Our hope is that by following the example of the human visual system, the robot's behavior will be easily understood because it is analogous to the behavior of a human in similar circumstances. For example, when an anthropomorphic robot moves its eyes and neck to orient toward an object, an observer can effortlessly conclude that the robot has become interested in that object (as discussed in chapter 6). These traits not only lead to behavior that is easy to understand, but also allow the robot's behavior to fit into the social norms that the person expects.

Another advantage is robustness. A system that integrates action, perception, attention, and other cognitive capabilities can be more flexible and reliable than a system that focuses on only one of these aspects. Adding additional perceptual capabilities and additional constraints between behavioral and perceptual modules can increase the relevance of behaviors while limiting the computational requirements. For example, in isolation, two difficult problems for a visual tracking system are knowing what to track and knowing when to switch to a new target. These problems can be simplified by combining the tracker with a visual attention system that can identify objects that are behaviorally relevant and worth tracking. In addition, the tracking system benefits the attention system by maintaining the object of interest in the center of the visual field. This simplifies the computation necessary to implement behavioral habituation. These two modules work in concert to compensate for the deficiencies of the other and to limit the required computation in each.

Using the human visual system as a model, a set of design criteria for Kismet's visual system can be specified. These criteria not only address performance issues, but aesthetic issues as well. The importance of functional aesthetics for performance as well as social constraints has been discussed in depth in chapter 5.

Similar visual morphology   Special attention has been paid to balancing the functional and aesthetic aspects of Kismet's camera configuration. From a functional perspective, the cameras in Kismet's eyes have high acuity but a narrow field of view. Between the eyes, there are two unobtrusive central cameras fixed with respect to the head, each with a wider field of view but correspondingly lower acuity.

The reason for this mixture of cameras is that typical visual tasks require both high acuity and a wide field of view. High acuity is needed for recognition tasks and for controlling precise visually guided motor movements. A wide field of view is needed for search tasks, for tracking multiple objects, compensating for involuntary ego-motion, etc. As described earlier, a common trade-off found in biological systems is to sample part of the visual field at a high resolution to support the first set of tasks, and to sample the rest of the field at an adequate level to support the second set. This is seen in animals with foveal vision, such as humans, where the density of photoreceptors is highest at the center and falls off dramatically towards the periphery. This can be implemented by using specially designed imaging hardware (van der Spiegel et al., 1989; Kuniyoshi et al., 1995), space-variant image sampling (Bernardino & Santos-Victor, 1999), or by using multiple cameras with different fields of view, as with Kismet.

Aesthetically, Kismet's big blue eyes are no accident. The cosmetic eyeballs envelop the fovea cameras and greatly enhance the readability of Kismet's gaze. The pair of minimally obtrusive wide field of view cameras that move with respect to the head are no accident, either. I did not want their size or movement to distract from Kismet's gaze. By keeping these other cameras inconspicuous, a person's attention is drawn to Kismet's eyes where powerful social cues are conveyed.

Similar visual perception   For robots and humans to interact meaningfully, it is important that they understand each other enough to be able to shape each other's behavior. This has several implications. One of the most basic is that robot and human should have at least some overlapping perceptual abilities (see chapters 5, 6, and 7). Otherwise, they can have little idea of what the other is sensing and responding to. Similarity of perception requires more than similarity of Sensors, however. Not all sensed stimuli are equally behaviorally relevant. It is important that both human and robot find the same types of stimuli salient in similar conditions. For this reason, Kismet is designed to have a set of perceptual biases based on the human pre-attentive visual system. I have discussed this issue at length in chapter 6.

Similar visual attention   Visual perception requires high bandwidth and is computationally demanding. In the early stages of human vision, the entire visual field is processed in parallel. Later computational steps are applied much more selectively, so that behaviorally relevant parts of the visual field can be processed in greater detail. This mechanism of visual attention is just as important for robots as it is for humans, from the same considerations of resource allocation. The existence of visual attention is also key to satisfying the expectations of humans concerning what can and cannot be perceived visually. Recall that chapter 6 presented the implementation of Kismet's context-dependent attention system that goes some way toward this.

Similar eye movements   Kismet's visual behaviors address both functional and social issues. From a functional perspective, Kismet uses a set of human-like visual behaviors that allow it to process the visual scene in a robust and efficient manner. These include saccadic eye movements, smooth pursuit, target tracking, gaze fixation, and ballistic head-eye orientation to target. We have also implemented two visual responses that very roughly approximate the function of the VOR (however, the current implementation does not employ a vestibular system), and the OKN. Due to human sensitivity to gaze, it is absolutely imperative that Kismet's eye movements look natural. Quite frankly, people find it disturbing if they move in a non-human manner.

Kismet's rich visual behavior can be conceptualized on those four levels presented in chapter 9 (namely, the social level, the behavior level, the skills level, and the primitives level). We have already argued how human-like visual behaviors have high communicative value in different social contexts. Higher levels of motor control address these social issues by coordinating the basic visual motor primitives (saccade, smooth pursuit, etc.) in a socially appropriate manner. We describe these levels in detail below, starting at the lowest level (the oculo-motor level) and progressing to the highest level where I discuss the social constraints of animate vision.

12.3   The Oculo-Motor System

The implementation of an oculo-motor system is an approximation of the human system. The system has been a large-scale engineering effort with substantial contributions by Brian Scassellati and Paul Fitzpatrick (Breazeal & Scassellati, 1999a; Breazeal et al., 2000). The motor primitives are organized around the needs of higher levels, such as maintaining and breaking mutual regard, performing visual search, etc. Since our motor primitives are tightly bound to visual attention, I will first briefly survey their sensory component.

Low-Level Visual Perception

Recall from chapter 5 and chapter 6, a variety of perceptual feature detectors have been implemented that are particularly relevant to interacting with people and objects. These include low-level feature detectors attuned to quickly moving objects, highly saturated color, and colors representative of skin tones. Looming and threatening objects are also detected pre-attentively, to facilitate a fast reflexive withdrawal (see chapter 6).

Visual Attention

Also presented in chapter 6, Wolfe's model of human visual search has been implemented and then supplemented to operate in conjunction with time-varying goals, with moving cameras, and to address the issue of habituation. This combination of top-down and bottom-up contributions allows the robot to select regions that are visually salient and behaviorally relevant. It then directs its computational and behavioral resources towards those regions. The attention system runs all the time, even when it is not controlling gaze, since it determines the perceptual input to which the motivational and behavioral systems respond.

In the presence of objects of similar salience, it is useful be able to commit attention to one object for a period of time. This gives time for post-attentive processing to be carried out on the object, and for downstream processes to organize themselves around the object. As soon as a decision is made that the object is not behaviorally relevant (for example, it may lack eyes, which are searched for post-attentively), attention can be withdrawn from it and visual search may continue. Committing to an object is also useful for behaviors that need to be atomically applied to a target (for example, the calling behavior where the robot needs to stay looking at the person it is trying to engage).

To allow such commitment, the attention system is augmented with a tracker. The tracker follows a target in the wide visual field, using simple correlation between successive frames. Changes in the tracker target are often reflected in movements of the robot's eyes, unless this is behaviorally inappropriate. If the tracker loses the target, it has a very good chance of being able to reacquire it from the attention system. Figure 12.3 shows the tracker in operation, which also can be seen in the CD-ROM's sixth demonstration, "Visual Behaviors."

Figure 12.3
Behavior of the tracker. Frames are taken at one-second intervals. The white squares indicate the position of the target. The target is not centered in the images since they were taken from a camera fixed with respect to the head, rather than gaze direction. On the third row, the face slips away from the tracker, but it is immediately reacquired through the attention system.

Post-Attentive Processing

Once the attention system has selected regions of the visual field that are potentially behaviorally relevant, more intensive computation can be applied to these regions than could be applied across the whole field. Searching for eyes is one such task. Locating eyes is important to us for engaging in eye contact, and as a reference point for interpreting facial movements and expressions. We currently search for eyes after the robot directs its gaze to a locus of attention, so that a relatively high resolution image of the area being searched is available from the foveal cameras (recall chapter 6). Once the target of interest has been selected, its proximity to the robot is estimated using a stereo match between the two central wide cameras (also discussed in chapter 6). Proximity is important for interaction as things closer to the robot should be of greater interest. It's also useful for interaction at a distance, such as a person standing too far away for face-to-face interaction but close enough to be beckoned closer. Clearly the relevant behavior (calling or playing) is dependent on the proximity of the human to the robot.

Eye Movements

Figure 12.4 shows the organization of Kismet's eye/neck motor control. Kismet's eyes periodically saccade to new targets chosen by an attention system, tracking them smoothly if they move and the robot wishes to engage them. Vergence eye movements are more challenging to implement in a social setting, since errors in disjunctive eye movements can give the eyes a disturbing appearance of moving independently. Errors in conjunctive movements have a much smaller impact on an observer, since the eyes clearly move in lock-step. A crude approximation of the opto-kinetic reflex is rolled into the implementation of smooth pursuit. Kismet uses an efferent copy mechanism to compensate the eyes for movements of the head.

The attention system operates on the view from the central camera. A transformation is needed to convert pixel coordinates in images from this camera into position set-points for the eye motors. This transformation in general requires the distance to the target to be known, since objects in many locations will project to the same point in a single image (see figure 12.5). Distance estimates are often noisy, which is problematic if the goal is to center the target exactly in the eyes. In practice, it is usually enough to get the target within the field of view of the foveal cameras in the eyes. Clearly, the narrower the field of view of these cameras, the more accurately the distance to the object needs to be known. Other crucial factors are the distance between the wide and foveal cameras, and the closest distance at which the robot will need to interact with objects. These constraints are determined by the physical distribution of Kismet's cameras and the choice of lenses. The central location of the wide camera places it as close as possible to the foveal cameras. It also has the advantage that moving the head to center a target in the central camera will in fact truly orient the head

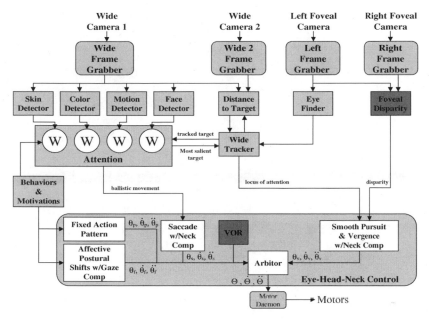

Figure 12.4
Organization of Kismet's eye/neck motor control. Many cross-level influences have been omitted. The modules in darkest gray are not active in the results presented in this chapter.

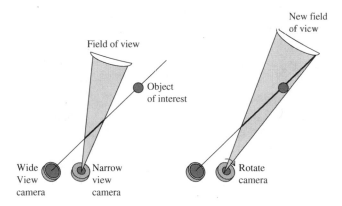

Figure 12.5
Without distance information, knowing the position of a target in the wide camera only identifies a ray along which the object must lie, and does not uniquely identify its location. If the cameras are close to each other (relative to the closest distance the object is expected to be at) the foveal cameras can be rotated to bring the object within their narrow field of view without needing an accurate estimate of its distance. If the cameras are far apart, or the field of view is very narrow, the minimum distance at which the object can be becomes large. The former solution is used in Kismet.

toward that target. For cameras in other locations, accuracy of orientation would be limited by the accuracy of the distance measurement.

Higher-level influences modulate the movement of the neck and eyes in a number of ways. As already discussed, modifications to weights in the attention system translate to changes of the locus of attention about which eye movements are organized. The overall posture of the robot can be controlled in terms of a three-dimensional affective space (chapter 10). The regime used to control the eyes and neck is available as a set of primitives to higher-level modules. Regimes include low-commitment search, high-commitment engagement, avoidance, sustained gaze, and deliberate gaze breaking. The primitive percepts generated by this level include a characterization of the most salient regions of the image in terms of the feature maps, an extended characterization of the tracked region in terms of the results of post-attentive processing (eye detection, distance estimation), and signals related to undesired conditions, such as a looming object, or an object moving at speeds the tracker finds difficult to keep up with.

12.4 Visual Motor Skills

Recall from chapter 9, given the current task (as dictated by the behavior system), the motor skills level is responsible for figuring out how to move the actuators to carry out the stated goal. Often this requires coordination between multiple motor modalities (speech, body posture, facial display, and gaze control).

The motor skills level interacts with both the behavior level above and the primitives level below. Requests for visual skills (each implemented as a FSM) typically originate from the behavior system. During turn-taking, for instance, the behavior system requests different visual primitives depending upon when the robot is trying to relinquish the floor (tending to make eye contact with the human) or to reacquire the floor (tending to avert gaze to break eye contact). Another example is the searching behavior. Here, the search FSM alternates ballistic orienting movements of the head and eyes to scan the scene with periods of gaze fixation to lock on the desired salient stimulus. The phases of ballistic orientations with fixations are appropriately timed to allow the perceptual flow of information to reach the behavior releasers and stop the search behavior when the desired stimulus is found. If the timing were too rapid, the searching behavior would never stop.

12.5 Visual Behavior

The behavior level is responsible for establishing the current task for the robot through arbitrating among Kismet's goal-achieving behaviors. By doing so, the observed behavior should be relevant, appropriately persistent, and opportunistic. The details of how this is accomplished are presented in chapter 9 and can be seen in figure 9.7. Both the current environmental conditions (as characterized by high-level perceptual releasers), as well as motivational factors such as `emotion` processes and homeostatic regulation processes, contribute to this decision process.

Interaction of the behavior level with the social level occurs through the world, as determined by the nature of the interaction between Kismet and the human. As the human responds to Kismet, the robot's perceptual conditions change. This can activate a different behavior, whose goal is physically carried out by the underlying motor systems. The human observes the robot's ensuing response and shapes their reply accordingly.

Interaction of the behavior level with the motor skills level also occurs through the world. For instance, if Kismet is looking for a bright toy, then the `seek-toy` behavior is active. This task is passed to the underlying motor skill that carries out the search. The act of scanning the environment brings new perceptions to Kismet's field of view. If a toy is found, then the `seek-toy` behavior is successful and released. At this point, the perceptual conditions for engaging the toy are relevant and the `engage-toy` behaviors become active. Consequently, another set of motor skills become active in order to track and smoothly pursue the toy. This indicates a significantly higher level of interest and engagement.

12.6   Visual Behavior and Social Interplay

The social level explicitly deals with issues pertaining to having a human in the interaction loop. As discussed previously, Kismet's eye movements have high communicative value. Its gaze direction indicates the locus of attention. Knowing the robot's locus of attention reveals what the robot currently considers to be behaviorally relevant. The robot's degree of engagement can also be conveyed to communicate how strongly the robot's behavior is organized around what it is currently looking at. If the robot's eyes flick about from place to place without resting, that indicates a low level of engagement, appropriate to a visual search behavior. Prolonged fixation with smooth pursuit and orientation of the head towards the target conveys a much greater level of engagement, suggesting that the robot's behavior is very strongly organized about the locus of attention. Eye movements are particularly potent during social interactions, such as conversational turn-taking, where making and breaking eye contact plays a role in regulating the exchange. As discussed previously, I have modeled Kismet's eye movements after humans, so that Kismet's gaze may have similar communicative value.

Eye movements are the most obvious and direct motor actions that support visual perception, but they are by no means the only ones. Postural shifts and fixed action patterns involving the entire robot also have an important role. Kismet has a number of coordinated motor actions designed to deal with various limitations of Kismet's visual perception

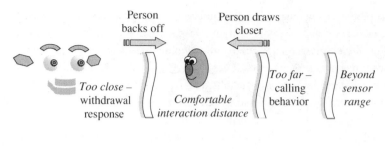

Person
backs off

Person draws
closer

Too close –
withdrawal
response

*Comfortable
interaction distance*

*Too far –
calling
behavior*

*Beyond
sensor
range*

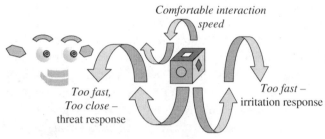

*Comfortable interaction
speed*

*Too fast,
Too close –*
threat response

Too fast –
irritation response

Figure 12.6
Regulating interaction via social amplification.

(see figure 12.6). For example, if a person is visible, but is too distant for their face to be imaged at adequate resolution, Kismet engages in a calling behavior to summon the person closer. People who come too close to the robot also cause difficulties for the cameras with narrow fields of view, since only a small part of a face may be visible. In this circumstance, a withdrawal response is invoked, where Kismet draws back physically from the person. This behavior, by itself, aids the cameras somewhat by increasing the distance between Kismet and the human. But the behavior can have a secondary and greater effect through *social amplification*—for a human close to Kismet, a withdrawal response is a strong social cue to back away, since it is analogous to the human response to invasions of "personal space." Hence, the consequence of Kismet's physical movement aids vision to some extent, but the social interpretation of this movement modulates the person's behavior in a strongly beneficial way for the robot. (The CD-ROM's fifth demonstration, "Social Amplification," illustrates this.)

Similar kinds of behavior can be used to support the visual perception of objects. If an object is too close, Kismet can lean away from it; if it is too far away, Kismet can crane its neck toward it. Again, in a social context, such actions have power beyond their immediate physical consequences. A human, reading intent into the robot's actions, may amplify those actions. For example, neck-craning towards a toy may be interpreted as interest in that toy, resulting in the human bringing the toy closer to the robot.

Another limitation of the visual system is how quickly it can track moving objects. If objects or people move at excessive speeds, Kismet has difficulty tracking them continuously. To bias people away from excessively boisterous behavior in their own movements or in the movement of objects they manipulate, Kismet shows irritation when its tracker is at the limits of its ability. These limits are either physical (the maximum rate at which the eyes and neck move), or computational (the maximum displacement per frame from the cameras over which a target is searched for).

Such regulatory mechanisms play roles in more complex social interactions, such as conversational turn-taking. Here control of gaze direction is important for regulating conversation rate (Cassell, 1999a). In general, people are likely to glance aside when they begin their turn, and make eye contact when they are prepared to relinquish their turn and await a response. Blinks occur most frequently at the end of an utterance. These and other cues allow Kismet to influence the flow of conversation to the advantage of its auditory processing. Kismet, however, does not perceive these gaze cues when used by others. Here, the visual-motor system is driven by the requirements of a nominally unrelated sensory modality, just as behaviors that seem completely orthogonal to vision (such as ear-wiggling during the calling behavior to attract a person's attention) are nevertheless recruited for the purposes of regulation.

These mechanisms also help protect the robot. Objects that suddenly appear close to the robot trigger a looming reflex, causing the robot to quickly withdraw and appear startled. If the event is repeated, the response quickly habituates and the robot simply appears annoyed, since its best strategy for ending these repetitions is to clearly signal that they are undesirable. Similarly, rapidly moving objects close to the robot are "threatening" and trigger an escape response.

These mechanisms are all designed to elicit natural and intuitive responses from humans, without any special training. But even without these carefully crafted mechanisms, it is often clear to a human when Kismet's perception is failing, and what corrective action would help. This is because the robot's perception is reflected in familiar behavior. Inferences made based on our human preconceptions are actually likely to work.

12.7 Evidence of Social Amplification

To evaluate the social implications of Kismet's behavior, we invited a few people to interact with the robot in a free-form exchange. There were four subjects in the study, two males (one adult and one child) and two females (both adults). They ranged in age from twelve to twenty-eight. None of the subjects were affiliated with MIT. All had substantial experience with computers. None of the subjects had any prior experience with Kismet. The child had prior experience with a variety of interactive toys. Each subject interacted with the robot for twenty to thirty minutes. All exchanges were video recorded for further analysis.

For the purposes of this chapter, I analyzed the video for evidence of social amplification. Namely, did people read Kismet's cues and did they respond to them in a manner that benefited the robot's perceptual processing or its behavior? I found several classes of interactions where the robot displayed social cues and successfully regulated the exchange.

Establishing a Personal Space

The strongest evidence of social amplification was apparent in cases where people came within very close proximity of Kismet. In numerous instances the subjects would bring their face very close to the robot's face. The robot would withdraw, shrinking backwards, perhaps with an annoyed expression on its face. In some cases the robot would also issue a vocalization with an expression of disgust. In one instance, the subject accidentally came too close and the robot withdrew without exhibiting any signs of annoyance. The subject immediately queried, "Am I too close to you? I can back up," and moved back to put a bit more space between himself and the robot. In another instance, a different subject intentionally put his face very close to the robot's face to explore the response. The robot withdrew while displaying full annoyance in both face and voice. The subject immediately pushed backwards, rolling the chair across the floor to put about an additional three feet between himself and the robot, and promptly apologized to the robot. (Similar events can be viewed on the sixth CD-ROM demonstration, "Visual Behavior.")

Overall, across different subjects, the robot successfully established a personal space. As discussed in the previous section, this benefits the robot's visual processing by keeping people at a distance where the visual system can detect eyes more robustly. This behavioral response was added to the robot's repertoire because previous interactions with naive subjects illustrated the robot was not granted any personal space. This can be attributed to "baby movements" where people tend to get extremely close to infants, for instance.

Luring People to a Good Interaction Distance

People seem responsive to Kismet's calling behavior. When a person is close enough for the robot to perceive his/her presense, but too far away for face-to-face exchange, the robot issues this social display to bring the person closer (see chapter 10). The most distinguishing features of the display are craning the neck forward in the direction of the person, wiggling the ears with large amplitude, and vocalizing with an excited affect. The function of the display is to lure people into an interaction distance that benefits the vision system. This behavior is not often witnessed as most subjects simply pull up a chair in front of the robot and remain seated at a typical face-to-face interaction distance (one example can be viewed on the fifth CD-ROM demonstration, "Social Amplification").

The youngest subject took the liberty of exploring different interaction ranges, however. Over the course of about fifteen minutes he would alternately approach the robot to a normal face-to-face distance, move very close to the robot (invading its personal space), and backing away from the robot. Upon the first appearance of the calling response, the experimenter queried the subject about the robot's behavior. The subject interpreted the display as the robot wanting to play, and he approached the robot. At the end of the subject's investigation, the experimenter queried him about the further interaction distances. The subject responded that when he was further from Kismet, the robot would lean forward. He also noted that the robot had a harder time looking at his face when he was farther back. In general, he interpreted the leaning behavior as the robot's attempt to initiate an exchange with him. I have noticed from earlier interactions (with other people unfamiliar with the robot) that a few people have not immediately understood this display as a calling behavior. The display is flamboyant enough, however, to arouse their interest to approach the robot.

Inferring the Level of Engagement

People seem to have a very good sense of when the robot is interested in a particular stimulus or not. By observing the robot's visual behavior, people can infer the robot's level of engagement toward a particular stimulus and generally try to be accommodating. This benefits the robot by bringing it into contact with the desired stimulus. I have already discussed an aspect of this in chapter 6 with respect to directing the robot's attention. Sometimes, however, the robot requires a different stimulus than the one being presented. For instance, the subject may be presenting the robot with a brightly colored toy, but the robot is actively trying to satiate its `social-drive` and searching for something skin-toned. As the subject tries to direct the robot's attention to the toy, the motion is enough to have the robot glance toward it (during the hold-gaze portion of the search behavior). Not being the desired stimulus, however, the robot moves its head and eyes to look in another direction. The subject often responds something akin to, "You don't want this? Ok, how about this toy?" as he/she attempts to get the robot interested in a different toy. Most likely the robot settles its gaze on the person's face fairly quickly. Noticing that the robot is more interested in them than the toy, they will begin to engage the robot vocally.

12.8   Limitations and Extensions

The data from these interactions is encouraging, but more formal studies with a larger number of subjects should be carried out. Whenever introducing a new person to Kismet, there is typically a getting acquainted period of five to ten minutes. During this time, the person gets a sense of the robot's behavioral repertoire and its limitations. As they notice "hiccups" in the interaction, they begin to more closely read the robot's cues and adapt their behavior. Great care was taken in designing these cues so that people intuitively understand the conditions under which they are elicited and what function they serve. Evidence shows that people readily and willingly read these cues to adapt their behavior in a manner that benefits the robot.

Unfortunately, twenty to thirty minutes is insufficient time to observe all of Kismet's cues, or to observe all the different types of interactions that Kismet has been designed to handle. For each subject, only a subset of these interactions were encountered. Often there is a core set of interactions that most people readily engage in with the robot (such as vocal exchanges and using a toy to play with the robot). The other interactions are more serendipitous (such as exploring the robot's interaction at a distance). People are also constrained by social norms. They rarely do anything that would be threatening or intentionally annoying to the robot. Thus, I have not witnessed how naive subjects interpret the robot's protective responses (such as its fear and escape response).

Extending Oculo-Motor Primitives

There are a couple of extensions that should be made to the oculo-motor system. The vestibulo-ocular reflex (VOR) is only an approximation of the human counterpart. Largely this is because the robot did not have the equivalent of a vestibular system. However, this issue has been rectified. Kismet now has a three DoF inertial sensor that measures head orientation (as the vestibular system does for people). My group has already developed VOR code for other robots, so porting the code to Kismet will happen soon. The second extension is to add vergence movements. It is very tricky to implement vergence on a robot like Kismet, because small corrections of each eye give the robot's gaze a chameleon-esque quality that is disturbing for people to look at. Computing a stereo map from the central wide field of view cameras would provide the foveal cameras with a good depth estimate, which could then be used to verge the eyes on the desired target. Since Kismet's eyes are fairly far apart, there is no attempt to exactly center the target with each fovea camera as this gives the robot a cross-eyed appearance even for objects that are nearby, but not invading the robot's personal space. Hence, there are many aesthetic issues that must be addressed as we implement these visual capabilities so as not to offend the human who interacts with Kismet.

Improving Social Responsiveness

There are several ways in which Kismet's social responsiveness can be immediately improved. Many of these relate to the robot's limited perceptual abilities. Some of these are issues of robustness, of latency, or of both.

Kismet's interaction performance at a distance needs to be improved. When a person is within perceptual range, the robot should make a compelling attempt to bring the person closer. The believability of the robot's behavior is closely tied to how well it can maintain mutual regard with that person. This requires that the robot be more robust in detecting people and their faces at a distance. The difference between having Kismet issue the calling display while looking at a person's face versus looking away from the person is enormous. I find that a person will not interpret the calling display as a request for engagement unless the robot is looking at their face when performing the display. It appears that the robot's gaze direction functions as a sort of social pointer—it says, "I'm directing this request and sending this message to you." For compelling social behavior, it's very important to get gaze direction right.

The perceptual performance can be improved by employing multi-resolution sampling on the camera images. Regions of the wide field of view that indicate the presence of skin-tone could be sampled at a higher resolution to see if that patch corresponds to a person. This requires another stage of processing that is not in the current implementation. If promising, the foveal camera could then be directed to look at that region to see if it can detect a face. Currently the foveal camera only searches for eyes, but at these distances the person's face is too small to reliably detect eyes. A face detector would have to be written for the foveal camera. If the presence of a face has been confirmed, then this target should be passed to the attention system to maintain this region as the target for the duration of the calling behavior. Other improvements to the visual system were discussed in chapter 6. These would also benefit interaction with humans.

12.9   Summary

Motor control for a social robot poses challenges beyond issues of stability and accuracy. Motor actions will be perceived by human observers as semantically rich, regardless of whether the imputed meaning is intended or not. This can be a powerful resource for facilitating natural interactions between robot and human, and places constraints on the robot's physical appearance and movement. It allows the robot to be readable—to make its behavioral intent and motivational state transparent at an intuitive level to those it interacts with. It allows the robot to regulate its interactions to suit its perceptual and motor capabilities, again in an intuitive way with which humans naturally co-operate. And it gives the robot leverage over the world that extends far beyond its physical competence, through social amplification of its perceived intent. If properly designed, the robot's visual behaviors can be matched to human expectations and allow both robot and human to participate in natural and intuitive social interactions.

I have found that different subjects have different personalities and different interaction styles. Some people read Kismet's cues more readily than others. Some people take longer to adapt their behavior to the robot. For the small number of subjects, I have found that people do intuitively and naturally adapt their behavior to the robot. They tune themselves to the robot in a manner that benefits the robot's computational limitations and improves the quality of the exchange. As is evident in the video, they enjoy playing with the robot. They express fondness of Kismet. They tell Kismet about their day and about personal experiences. They treat Kismet with politeness and consideration (often apologizing if they have irritated the robot). They often ask the robot what it likes, what it wants, or how it feels in an attempt to please it. The interaction takes place on a physical, social, and affective level. In so many ways, they treat Kismet as if it were a socially aware, living creature.

13 Grand Challenges of Building Sociable Robots

Human beings are a social species of extraordinary ability. Overcoming social challenges has played a significant role in our evolution. Interacting socially with others is critical for our development, our education, and our day-to-day existence as members of a greater society. Our sociability touches upon the most human of qualities: personality, identity, emotions, empathy, loyalty, friendship, and more. If we are to ever understand human intelligence, human nature, and human identity, we cannot ignore our sociality.

The directions and approaches presented in this book are inspired by human social intelligence. Certainly, my experiences and efforts in trying to capture a few aspects of even the simplest form of human social behavior (that of a human infant) has been humbling, to say the least. In the end, it has deepened my appreciation of human abilities. Through the process of building a sociable robot, from Kismet and beyond, I hope to achieve a deeper understanding of this fascinating subject.

In this chapter I recap the significant contributions of this body of work with Kismet, and then look to the future. I outline some grand challenge problems for building a robot whose social intelligence might someday rival our own. The field of sociable robotics is nascent, and much work remains to be done. I do not claim that this is a complete treatment. Instead, these challenge problems will be subject to revision over time, as new challenges are encountered and old challenges are resolved. My work with Kismet touches on some of these grand challenge problems. A growing number of researchers have begun to address others. I highlight a few of these efforts in this chapter, concentrating on work with autonomous robots.

The preceding chapters give an in-depth presentation of Kismet's physical design and the design of its synthetic nervous system. A series of issues that have been found important when designing autonomous robots that engage humans in natural, intuitive, and social interaction have been outlined. Some of these issues pertain to the physical design of the robot: its aesthetics, its sensory configuration, and its degrees of freedom. Kismet was designed according to these principles.

Other issues pertain to the design of the synthetic nervous system. To address these computational issues, this book presents a framework that encompasses the architecture, the mechanisms, the representations, and the levels of control for building a sociable machine. I have emphasized how designing for a human in the loop profoundly impacts how one thinks about the robot control problem, largely because robot's actions have social consequences that extend far beyond the immediate physical act. Hence, one must carefully consider the social constraints imposed on the robot's observable behavior. The designer can use this to benefit the quality of interaction between robot and human, however, as illustrated in the numerous ways Kismet proactively regulates its interaction with the human so that the interaction is appropriate for both partners. The process of social amplification is a prime example.

In an effort to make the robot's behavior readable, believable, and well-matched to the human's social expectations and behavior, several theories, models, and concepts from psychology, social development, ethology, and evolutionary perspectives are incorporated into the design of the synthetic nervous system. I highlighted how each system addresses important issues to support natural and intuitive communication with a human and how special attention was paid to designing the infrastructure into the synthetic nervous system to support socially situated learning.

These diverse capabilities are integrated into a single robot situated within a social environment. The performance of the human-robot system with numerous studies with human subjects has been evaluated (Breazeal, 2002). Below I summarize the findings as they pertain to the key design issues and evaluation criteria outlined in chapter 4.

13.1 Summary of Key Design Issues

Through these studies with human subjects, I have found that Kismet addresses the key design issues in rich and interesting ways. By going through each design issue, I recap the different ways in which Kismet meets the four evaluation criteria. Recall from chapter 4, these criteria are:

• Do people intuitively read and naturally respond to Kismet's social cues?

• Can Kismet perceive and appropriately respond to these naturally offered cues?

• Does the human adapt to the robot, and the robot adapt to the human, in a way that benefits the interaction?

• Does Kismet readily elicit scaffolding interactions from the human that could be used to benefit learning?

Real-time performance Kismet successfully maintains interactive rates in all of its systems to dynamically engage a human. I discussed the performance latencies of several systems including visual and auditory perception, visual attention, lip synchronization, and turn-taking behavior during proto-dialogue. Although each of these systems does not perform at adult human rates, they operate fast enough to allow a human engage the robot comfortably. The robot provides important expressive feedback to the human that they intuitively use to entrain to the robot's level of performance.

Establishment of appropriate social expectations Great care has been taken in designing Kismet's physical appearance, its sensory apparatus, its mechanical specification, and its observable behavior (motor acts and vocal acts) to establish a robot-human relationship that follows the infant-caregiver metaphor. Following the baby-scheme of Eibl-Eiblsfeldt, Kismet's appearance encourages people to treat it as if it were a very young child or infant. Kismet has been given a child-like voice and it babbles in its own characteristic manner.

Female subjects are willing to use exaggerated prosody when talking to Kismet, characteristic of motherese. Both male and female subjects tend to sit directly in front of and close to Kismet, facing it the majority of the time. When engaging Kismet in proto-dialogue, they tend to slow down, use shorter phrases, and wait longer for Kismet's response. Some subjects use exaggerated facial expressions. All these behaviors are characteristic of interacting with very young animals (e.g., puppies) or infants.

Self-motivated interaction   Kismet exhibits self-motivated and proactive behavior. Kismet is in a never-ending cycle of satiating its `drives`. As a result, the stimuli it actively seeks out (people-like things versus toy-like things) changes over time. The first level of the behavior system acts to seek out the desired stimulus when it is not present, to engage it when it has been found, and to avoid it if it is behaving in an offensive or threatening manner. The gains of the attention system are dynamically adjusted over time to facilitate this process. Kismet can take the initiative in establishing an interaction. For instance, if Kismet is in the process of satiating its `social-drive`, it will call to a person who is present but slightly beyond face-to-face interaction distance.

Regulation of interactions   Kismet is well-versed in regulating its interactions with the caregiver. It has several mechanisms for accomplishing this, each for different kinds of interactions. They all serve to slow the human down to an interaction rate that is within the comfortable limits of Kismet's perceptual, mechanical, and behavioral limitations. By doing so, the robot is neither overwhelmed nor under-stimulated by the interaction.

The robot has two regulatory systems that serve to maintain the robot in a state of "well-being." These are the emotive responses and the homeostatic regulatory mechanisms. The `drives` establish the desired stimulus and motivate the robot to seek it out and to engage it. The `emotions` are another set of mechanisms, with greater direct control over behavior and expression, that serve to bring the robot closer to desirable situations (`joy`, `interest`, even `sorrow`), and cause the robot to withdraw from or remove undesirable situations (`fear`, `anger`, or `disgust`). Which emotional response becomes active depends largely on the releasers, but also on the internal state of the robot. The behavioral strategy may involve a social cue to the caregiver (through facial expression and body posture) or a motor skill (such as the escape response). The use of social amplification to define a personal space is a good example of how social cues, that are a product of emotive responses, can be used to regulate the proximity of the human to the robot. It is also used to regulate the movement of toys when playing with the robot.

Kismet's turn-taking cues for regulating the rate of proto-dialogue is another case. Here, the interaction happens on a more tightly coupled temporal dynamic between human and robot. The mechanism originates from the behavior system instead of the emotion system. It employs communicative facial displays instead of emotive facial expressions. Our studies suggest that subjects read the robot's turn-taking cues to entrain to the robot. As a result, the proto-dialogue becomes smoother over time.

Readable social cues   Kismet is a very expressive robot. It can communicate "emotive" state and social cues to a human through face, gaze direction, body posture, and voice. Results from various forced-choice and similarity studies suggest that Kismet's emotive facial expressions and vocal expressions are readable. More importantly, several studies suggest that people readily read and correctly interpret Kismet's expressive cues when actively engaging the robot. I found that several interesting interactions arose between Kismet and female subjects when Kismet's ability to recognize vocal affective intent (for praise, prohibition, etc.) was combined with expressive feedback. The female subjects used Kismet's facial expression and body posture as a social cue to determine when Kismet "understood" their intent. The video of these interactions suggests evidence of affective feedback where the subject would issue an intent (say, an attentional bid), the robot would respond expressively (perking its ears, leaning forward, and rounding its lips), and then the subject would immediately respond in kind (perhaps by saying, "Oh!" or, "Ah!"). Several subjects appeared to empathize with the robot after issuing a prohibition—often reporting feeling guilty or bad for scolding the robot and making it "sad." For turn-taking interactions, after a period of entrainment, subjects appear to read the robot's social cues and hold their response until prompted by the robot. This allows for longer runs of clean turns before an interruption or delay occurs in the proto-dialogue.

Interpretation of human's social cues   I have presented two cases where the robot can read the human's social cues. The first is the ability to recognize praise, prohibition, soothing, and attentional bids from robot-directed speech. This could serve as an important teaching cue for reinforcing and shaping the robot's behavior. The second is the ability of humans to direct Kismet's attention using natural cues. This could play an important role in socially situated learning by giving the caregiver a way of showing Kismet what is important for the task, and for establishing a shared reference.

Competent behavior in a complex world   Kismet's behavior exhibits robustness, appropriateness, coherency, and flexibility when engaging a human in either physical play with a toy, in vocal exchanges, or affective interactions. It also exhibits appropriate persistence and reasonable opportunism when addressing its time-varying goals. These qualities arise from the interaction between the external environment with the internal dynamics of Kismet's synthetic nervous system. The behavior system is designed to address these issues on the task level, but the observable behavior is a product of the behavior system working in concert with the perceptual, attention, motivation, and motor systems. In chapter 9,

I conceptualized Kismet's behavior to be the product of interactions within and between four separate levels.

Believable behavior   Kismet exhibits compelling and life-like behavior. To promote this quality of behavior, the issues of audience perception and of biasing the robot's design towards believability, simplicity, and caricature over forced realism were addressed. A set of proto-social responses that are synthetic analogs of those believed to play an important role in launching infants into social exchanges with their caregivers have been implemented.

From video recordings of subjects interacting with Kismet, people do appear to treat Kismet as a very young, socially aware creature. They seem to treat the robot's expressive behaviors and vocalizations as meaningful responses to their own attempts at communication. The robot's prosody has enough variability that they answer Kismet's "questions," comment on Kismet's "statements," and react to Kismet's "exclamations." They ask Kismet about its thoughts and feelings, how its day is going, and they share their own personal experiences with the robot. These kinds of interactions are important to foster the social development of human infants. They could also play an important role in Kismet's social development as well.

13.2   Infrastructure for Socially Situated Learning

In the above discussion, I have taken care to relate these issues to socially situated learning. In previous work, my colleagues and I have posed these issues with respect to building humanoid robots that can imitate people (Breazeal & Scassellati, 2002). I quickly recap these issues below.

Knowing what's important   Determining what the robot should attend to is largely addressed by the design of the attention system. It is easy for people to direct Kismet's attention, as well as to confirm when the robot's attention has been successfully manipulated. People can also use their voice to arouse the robot through attentional bids. More work needs to be done, but this provides a solid foundation.

Recognizing progress   The robot is designed to have both internal mechanisms as well as external mechanisms for recognizing progress. The change in Kismet's internal state (the satiation of its `drives`, or the return to a slightly positive affective state) could be used as internal reinforcement signals for the robot. Other systems have used signals of this type for operant as well as classical conditioning of robotic or animated characters (Velasquez, 1998; Blumberg et al., 1996; Yoon et al., 2000). Kismet also has the ability to extract progress measures from the environment, through socially communicated praise, prohibition, and soothing. The underlying mechanism would actually be similar to the previous case, as the human is modulating the robot's affective state by communicating these intents. Eventually, this could be extended to having the robot recognize positive and negative facial expressions.

Recognizing success   The same mechanisms for recognizing progress could be used to recognize success. The ability for the caregiver to socially manipulate the robot's affective state has interesting implications for teaching the robot novel acts. The robot may not require an explicit representation of the desired goal nor a fully specified evaluation function before embarking upon learning the task. Instead, the caregiver could initially serve as the evaluation function for the robot, issuing praise, prohibition, and encouragement as she tries to shape the robot's behavior. It would be interesting if the robot could learn how to associate different affective states to the learning episode. Eventually, the robot may learn to associate the desired goal with positive affect—making that goal an explicitly represented goal within the robot instead of an implicitly represented goal through the social communication of affect. This kind of scenario could play an important part in socially transferring new goals from human to robot. Many details need to be worked out, but the kernel of the idea is intriguing.

Structured learning scenarios   Kismet has two strategies for establishing an appropriate learning environment. Both involve regulating the interaction with the human. The first takes place through the motivation system. The robot uses expressive feedback to indicate to the caregiver when it is either overwhelmed or under-stimulated. In time, this mechanism has been designed with the intent that homeostatic balance of the `drives` corresponds to a learning environment where the robot is slightly challenged but largely competent. The second form of regulation is turn-taking, which is implemented in the behavior system. Turn-taking is a cornerstone of human-style communication and tutelage. It forms the basis of interactive games and structured learning episodes. In the near future, these interaction dynamics could play an important role in socially situated learning for Kismet.

Quality instruction   Kismet provides the human with a wide assortment of expressive feedback through several different expressive channels. Currently, this is used to help entrain the human to the robot's level of competence, and to help the human maintain Kismet's "well-being" by providing the appropriate kinds of interactions at the appropriate times. This could also be used to intuitively help the human provide better quality instruction. Looks of puzzlement, nods or shakes of the head, and other gestures and expressions could be employed to elicit further assistance or clarification from the caregiver.

13.3   Grand Challenge Problems

The ultimate challenge for a sociable robot is to interact with humans as another person would and to be accepted as part of the human community. In chapter 1, I outlined several key ingredients of building robots that can engage people socially in a human-like way. This list was derived to support several key attributes of human sociality. These ingredients address the broader questions of building a socially intelligent artifact that is embodied and situated in the human social environment, that exhibits autonomy and life-like qualities to encourage people treat it as a socially aware entity, that perceives and understands human social behavior, that behaves in a way understandable to people in familiar social terms, that is self-aware and able to reflect upon its own mental states and those of others, that learns throughout its lifetime to increase its aptitude, and that continually adapts to new experiences to establish and maintain relationships with others. Some of the grand challenge problems are derived from these target areas. Other challenge problems address issues of evaluation, understanding the impact on the human who interacts with it, and understanding the impact on human society and culture.

Anima machina   As the term *anima machina* suggests, this grand challenge problem speaks to building a life-like robot.[1] This challenge encompasses both the construction of a robot that can manage its daily physical existence in human society, as well as the design of the synthetic nervous system that "breathes the life" into the machine.

With respect to the physical machine, overall robustness and longevity are important issues. Fortunately, advancements in power source technology, actuator design, sensors, computation hardware, and materials are under way. Improvements in power source size, weight, and lifetime are critical for robots that must carry their own batteries, fuel, etc. The ability for the robot to replenish its energy over time is also important. New actuator technologies have more muscle-like properties such as compliancy and energy storage (Pratt & Williamson, 1995; Robinson et al., 1999). Researchers are looking into better mechanisms that approximate the motion of complex rotational joints, such as shoulders (Okada et al., 2000), or that replicate the flexible movement of the spinal cord (Mizuuchi et al., 2000). Improvements in current sensor technologies, such as developing cameras that lend themselves to a more retina-like distribution of pixels (Kuniyoshi et al., 1995) or increasing the sensory repertoire to give a robot the sense of smell (Dickinson et al., 1999), are also under way. New materials are under investigation, such as gel-like actuators that might find interesting applications for synthetic skin (Otake et al., 1999). Cross-fertilization

---

1. Rod Brooks takes poetic license to convey this idea with the phrase "living, breathing robots."

of technologies from biomedical engineering may also present new possibilities for synthetic bodies.

Personality   This challenge problem concerns endowing sociable robots with rich personalities. This supports our tendency to anthropomorphize—to treat the robot as an individual with human-like qualities and mental states. By doing so, the robot is perceived as being enough like us that people can understand and predict the robot's behavior in familiar social terms. Furthermore, as the amount of social interaction with a technology increases, people want the technology to be believable (Bates, 1994). The success of cyber-pets such as PF Magic's Petz is a case in point. If a sociable robot had a compelling personality, there is reason to believe that people would be more willing to interact with it, would find the interaction more enjoyable, and would be more willing to establish a relationship with it. Animators have amassed many insights into how to convey the illusion of life (Thomas & Johnston, 1981). Researchers in the field of life-like characters apply many of these insights in an effort to design personality-rich interactive software agents. More recently, there is a growing appreciation and interest in giving autonomous robots compelling personalities in order to foster effective interactions with people. A growing number of commercial products target the toy and entertainment markets such as Tiger Electronic's Furby (a creature-like robot), Hasboro's My Real Baby (a robotic doll), and Sony's Aibo (a robotic dog).

Certainly, Kismet's personality is a crucial aspect of its design and has proven to be engaging to people. It is conveyed through aesthetic appearance, quality of movement, manner of expression, and child-like voice. Kismet's conveys a sense of sweetness, innocence, and curiosity. The robot communicates an "opinion," expressing approval and disapproval of how a person interacts with it. It goes through "mood swings," sometimes acting fussy, other times acting tolerant and content. This is an appropriate personality for Kismet given how we want people to interact with it, and given that Kismet is designed to explore those social learning scenarios that transpire between an infant and a caregiver.

Embodied discourse   This grand challenge problem targets a robot's ability to partake in natural human conversation as an equally proficient participant. To do so, the robot must be able to communicate with humans by using natural language that is also complemented by paralinguistic cues such as gestures, facial expressions, gaze direction, and prosodic variation.

One of the most advanced systems that tackles this challenge is Rea, a fully embodied animated-conversation-agent developed at the MIT Media Lab (Cassell et al., 2000). Rea is an expert in the domain of real estate, serving as a real-estate agent that humans can interact with to buy property. Rea supports conversational discourse and can sense human paralinguistic cues such as hand gestures and head pose. Rea communicates in kind, using variations in prosody, gesture, facial expression, and gaze direction. Our work with Kismet explores pre-linguistic communication where important paralinguistic cues such as gaze direction and facial expressions are used to perform key social skills, such as directing attention and regulating turn-taking during face-to-face interaction between the human and the robot. In related robotics work, an upper-torso humanoid robot called Robita can track the speaking turns of the participants during triadic conversations—i.e., between the robot and two other people (Matsusaka & Kobayashi, 1999). The robot has an expressionless face but is able to direct its attention to the appropriate person though head posture and gaze direction, and it can participate in simple verbal exchanges.

Personal recognition   This challenge problem concerns the recognition and representation of people as individuals who have distinct personalities and past experiences. To quote Dautenhahn (1998, p. 609), "humans are individuals and want to be treated as such." To establish and maintain relationships with people, a sociable robot must be able to identify and represent the people it already knows as well as add new people to its growing set of known acquaintances. Furthermore, a sociable robot must also be able to reflect upon past experiences with these individuals and take into account new experiences with them.

Toward this goal, a variety of technologies have been developed to recognize people in a variety of modalities such as visual face recognition, speaker identification, fingerprint analysis, retinal scans, and so forth. Chapter 1 mentions a number of different approaches for representing people and social events in order to understand and reason about social situations. For instance, story-based approaches have been explored by a number of researchers (Schank & Abelson, 1977; Bruner, 1991; Dautenhahn & Coles, 2001).

Theory of mind   This challenge problem addresses the issue of giving a robot the ability to understand people in social terms. Specifically, the ability for a robot to infer, represent, and reflect upon the intents, beliefs, and wishes of those it interacts with. Recall that chapter 1 discussed the theory of mind competence of humans, referring to our ability to attribute beliefs, goals, percepts, feelings, and desires to ourselves and to others. I outlined a number of different approaches being explored to give machines an analogous competence, such as modeling these mental states with explicit symbolic representations (Kinny et al., 1996), adapting psychological models for theory of mind from child development to robots (Scassellati, 2000a), employing a story-based approach based on scripts (Schank & Abelson, 1977), or through a process of biographic reconstruction as proposed in (Dautenhahn, 1999b).

Empathy   This challenge problem speaks to endowing a robot with the ability to infer, understand, and reflect upon the emotive states of others. Humans use empathy to know what others are feeling and to comprehend their positive and negative experiences. Brothers (1989, 1997) views empathy as a means of understanding and relating to others by wilfully changing one's own emotional and psychological state to mirror that of another. It is a fundamental human mechanism for establishing emotional communication with others. Siegel (1999) describes this state of communication as "feeling felt." More discussions of empathy in animals, humans, and robots can be found in Dautenhahn (1997).

Although work with Kismet does not directly address the question of empathy for a robot, it does explore an embodied approach to understanding the affective intent of others. Recall from chapter 7 that a human can induce an affective state in Kismet that roughly mirrors his or her own—either through praising, prohibiting, alerting, or soothing the robot. Kismet comes to "understand" the human's affective intent by adopting an appropriate affective state.

For technologies that must interact socially with humans, it is acknowledged that the ability to perceive, represent, and reason about the emotive states of others is important. For instance, the field of Affective Computing tries to measure and model the affective states of humans by using a variety of sensing technologies (Picard, 1997). Some of these sensors measure physiological signals such as skin conductance and heart rate. Other approaches analyze readily observable signals such as facial expressions (Hara, 1998) or variations in vocal quality and speech prosody (Nakatsu et al., 1999). Several symbolic AI systems, such as the Affective Reasoner by Elliot, adapt psychological models of human emotions in order to reason about people's emotional states in different circumstances (Elliot, 1992). Others explore computational models of emotions to improve the decision-making or learning processes in robots or software agents (Yoon et al., 2000; Velasquez, 1998; Canamero, 1997; Bates et al., 1992). Our work with Kismet explores how emotion-like processes can facilitate and foster social interaction between human and robot.

Autobiographic memory   This challenge problem concerns giving a robot the ability to represent and reflect upon its self and its past experiences. Chapter 1 discussed autobiographical memories in humans and their role in self-understanding. Dautenhahn (1998) introduces the notion of an *autobiographic agent* as "an embodied agent that dynamically reconstructs its individual 'history' (autobiography) during its lifetime."

Autobiographical memory develops during the lifetime of a human being and is socially constructed through interaction with others. The *social interaction hypothesis* states that children gradually learn the forms of how to talk about memory with others and thereby learn how to formulate their own memories as narratives (Nelson, 1993). Telling a reasonable autobiographical story to others involves constructing a plausible tale by weaving together not only the sequence of episodic events, but also one's goals, intentions, and motivations (Dautenhahn, 1999b). Cassell and Glos (1997) have shown how agent technologies could be used to help children develop their own autobiographical memory through creating and telling stories about themselves. A further discussion of narrative and autobiographical memory as applied to robots is provided in (Dautenhahn, 1999b).

Socially situated learning   This challenge problem concerns building a robot that can learn from humans in social scenarios. In chapters 1 and 2, I presented detailed discussions of the importance and advantages of socially situated learning for robots. The human social environment is always changing and is unpredictable. There are many social pressures requiring that a sociable robot learn throughout its lifetime. The robot must continuously learn about its self as new experiences shape its autobiographical memory. The robot also must learn continually from and adapt to new experiences that it shares with others to establish and maintain relationships. New skills and competencies can be acquired from others, either humans or other robots. This is a critical capability since the human social environment is too complex and variable to explicitly pre-program the robot with everything it will ever need to know.

In this book, I have motivated work with Kismet from the fact that humans naturally offer many different social cues to help others learn, and that a robot could also leverage from these social interactions to foster its own learning. Other researchers and I are exploring specific types of social learning, such as learning by imitation, to allow a human (or in some cases another robot) to transfer skills to a robot learner through direct demonstration (Schaal, 1997; Billard & Mataric, 2000; Ude et al., 2000; Breazeal & Scassellati, 2002).

Evaluation metrics   As the social intelligence of these robots increases, how will we evaluate them? Certainly, there are many aspects of a sociable robot that can be measured and quantified objectively, such as its ability to recognize faces, its accuracy of making eye contact, etc. Other aspects of the robot's performance, however, are inherently subjective (albeit quantifiable), such as the readability of its facial expressions, the intelligibility of its speech, the clarity of its gestures, etc. The evaluation of these subjective aspects of the design (such as the believability of the robot) varies with the person who interacts with it. A compelling personality to one person may be flat to another. The assessment of other attributes may follow demographic trends, showing strong correlations with age, gender, cultural background, education, and so forth. Establishing a set of evaluation criteria that unveils these correlations will be important for designing sociable robots that are well-matched to the people it interacts with.

If at some point in the future the sociability of these kinds of robots appears to rival our own, then empirical measures of performance may become extremely difficult to define, if not pointless. How do we empirically measure our ability to empathize with another, or another's degree of self-awareness? Ultimately what matters is how we treat them and how they treat us. What is the measure of a person, biological or synthetic?

Understanding the human in the loop   The question of how sociable robots should fit into society depends on how these technologies impact the people who interact with them. We must understand the human side of the equation. How will people interact with sociable robots? Will people accept them or fear them? How might this differ with age, gender, culture, etc.?

The idea of sociable robots coexisting with us in society is not new. Through novels and films, science fiction has shown us how wonderful or terrifying this could be. Sociable robots of this imagined sophistication do not exist, and it will be quite some time before they are realized. Their improvements will be incremental, driven by commercial applications as well as by the research community. Robotic toys, robot pets, and simple domestic robots already are being introduced into society as commercial products. As people interact with these technologies and try to integrate them into their daily lives, their attitudes and preferences will shape the design specification of these robots. Conversely, as the robots become more capable, people's opinions and expectations toward them will change, becoming more accepting of them, and perhaps becoming more reliant upon them. Sociable robots will grow and change with people, as people will grow and change with them.

The field of sociable robots is in its earliest stages. Research should target not only the engineering challenge of building socially intelligent robots, but also acquire a scientific understanding of the interaction between sociable robots and humans. As the field matures, understanding both sides of the human-robot equation will be critical to developing successful socially intelligent technologies that are well matched to the greater human community. Toward this goal, this book presents both the engineering aspects of building a sociable robot as well as the experimental aspects of how naive subjects interact with this kind of technology. Both endeavors have been critical to our research program.

Friendship   This challenge problem is perhaps the ultimate achievement in building a sociable robot. What would be required to build a robot that could be a genuine friend? We see examples of such robots in science fiction such as R2-D2 or C-3PO from the movie *Star Wars,* Lt. Commander Data from the television series *Star Trek: The Next Generation,* or the android Andrew from Isaac Asimov's short story *Bicentennial Man*. These robots exhibit some of our most prized human qualities. They have rich personalities, show compassion and kindness, can empathize and relate to their human counterparts, are loyal to their friends to the point of risking their own existence, behave with honor, and have a sense of character and honor.

Personhood   This challenge problem is not one of engineering, but one for society. What are the social implications of building a sociable machine that is capable of being a genuine friend? When is a machine no longer just a machine, but an intelligent and "living" entity that merits the same respect and consideration given to its biological counterparts? How will society treat a socially intelligent artifact that is not human but nonetheless seems to be a person? How we ultimately treat these machines, whether or not we grant them the status of personhood, will reflect upon ourselves and on society.

Foerst (1999; Foerst & Petersen, 1998) explores the questions of identity and personhood for humanoid robots, arguing that our answers ultimately reflect our views on the nature of being human and under what conditions we accept someone into the human community. These questions become increasingly more poignant as humans continue to integrate technologies into our lives and into our bodies in order to "improve" ourselves, augmenting and enhancing our biologically endowed capabilities. Eyeglasses and wristwatches are examples of how widely accepted these technological improvements can become. Modern medicine and biomedical engineering have developed robotic prosthetic limbs, artificial hearts, cochlear implants, and many other devices that allow us to move, see, hear, and live in ways that would otherwise not be possible. This trend will continue, with visionaries predicting that technology will eventually augment our brains to enhance our intellect (Kurzweil, 2000).

Consider a futuristic scenario where a person continues to replace his/her biological components with technologically enhanced counterparts. Taken to the limit, is there a point when he/she is no longer human? Is there a point where she/he is no longer a person? Foerst urges that this is not an empirical decision, that measurable performance criteria (such as measuring intelligence, physical capabilities, or even consciousness) should *not* be considered to assign personhood to an entity. The risk of excluding some humans from the human community is too great, and it is better to open the human community to robots (and perhaps some animals) rather than take this risk.

13.4   Reflections and Dreams

I hope that Kismet is a precursor to the socially intelligent robots of the future. Today, Kismet is the only autonomous robot that can engage humans in natural and intuitive interaction that is physical, affective, and social. At times, people interact with Kismet at a level that seems personal—sharing their thoughts, feelings and experiences with Kismet. They ask Kismet to share the same sorts of things with them.

After a three-year investment, we are in a unique position to study how people interact with sociable autonomous robots. The work with Kismet offers some promising results, but many more studies need to be performed to come to a deep understanding of how people interact with these technologies. Also, we are now in the position to study socially situated learning following the infant-caregiver metaphor. From its inception, this form of learning has been the motivation for building Kismet, and for building Kismet in its unique way.

In the near term, I am interested in emulating the process by which infants "learn to mean" (Halliday, 1975). Specifically, I am interested in investigating the role social interaction plays in how very young children (even African Grey parrots, as evidenced by the work of Pepperberg [1990]) learn the meaning their vocalizations have for others, and how to use

©2000 Peter Menzel/Robo sapiens

this knowledge to benefit its own behavior and communication. In short, I am interested in
having Kismet learn not only how to communicate, but also the function of communication
and how to use it pragmatically. There are so many different questions I want to explore in
this fascinating area of research. I hope I have succeeded in inspiring others to follow.

In the meantime, kids are growing up with robotic and digital pets such as Aibo, Furby,
Tomogotchi, Petz, and others soon to enter the toy market. Their experience with interactive
technologies is very different from that of their parents or grandparents. As the technology
improves and these children grow up, it will be interesting to see what is natural, intuitive,
and even expected of these interactive technologies. Sociable machines and other sociable
technologies may become a reality sooner than we think.

References

Adams, B., Breazeal, C., Brooks, R., Fitzpatrick, P., and Scassellati, B. (2000). "Humanoid robots: A new kind of tool," *IEEE Intelligent Systems* 15(4), 25–31.

Aldiss, B. (2001). *Supertoys Last All Summer Long: And Other Stories of Future Time,* Griffin Trade.

Ambrose, R., Aldridge, H., and Askew, S. (1999). NASA's Robonaut system, *in* "Proceedings of the Second International Symposium on Humanoid Robots (HURO99)," Tokyo, Japan, pp. 131–132.

Asimov, I. (1986). *Robot Dreams,* Berkley Books, New York, NY. A masterworks edition.

Ball, G., and Breese, J. (2000). Emotion and personality in a conversational agent, *in* J. Cassell, J. Sullivan, S. Prevost and E. Churchill, eds., "Embodied Conversational Agents," MIT Press, Cambridge, MA, pp. 189–219.

Ball, G., Ling, D., Kurlander, D., Miller, J., Pugh, D., Skelley, T., Stankosky, A., Thiel, D., Dantzich, M. V., and Wax, T. (1997). Lifelike computer characters: The Persona Project at Microsoft Research, *in* J. Bradshaw, ed., "Software Agents," MIT Press, Cambridge, MA.

Ballard, D. (1989). "Behavioral constraints on animate vision," *Image and Vision Computing* 7(1), 3–9.

Baron-Cohen, S. (1995). *Mindblindness: An Essay on Autism and Theory of Mind,* MIT Press, Cambridge, MA.

Barton, R., and Dunbar, R. (1997). Evolution of the social brain, *in* A. Whiten and R. Byrne, eds., "Machiavellian Intelligence II: Extensions and Evaluation," Cambridge University Press, Cambridge, UK, pp. 240–263.

Bates, J. (1994). "The role of emotion in believable characters," *Communications of the ACM* 37(7), 122–125.

Bates, J., Loyall, B., and Reilly, S. (1992). An architecture for action, emotion, and social behavior, Technical Report CMU-CS-92-144, Carnegie Mellon University, Pittsburgh, PA.

Bateson, M. (1979). The epigenesis of conversational interaction: a personal account of research development, *in* M. Bullowa, ed., "Before Speech: The Beginning of Interpersonal Communication," Cambridge University Press, Cambridge, UK, pp. 63–77.

Bernardino, A., and Santos-Victor, J. (1999). "Binocular visual tracking: Integration of perception and control," *IEEE Transactions on Robotics and Automation* 15(6), 1937–1958.

Billard, A., and Dautenhahn, K. (1997). Grounding Communication in Situated, Social Robots, Technical Report UMCS-97-9-1, University of Manchester.

Billard, A., and Dautenhahn, K. (1998). "Grounding communication in autonomous robots: An experimental study," *Robotics and Autonomous Systems* 1–2(24), 71–81.

Billard, A., and Dautenhahn, K. (2000). "Experiments in learning by imitation: Grounding and use of communication in robotic agents," *Adaptive Behavior* 7(3–4), 415–438.

Billard, A., and Mataric, M. (2000). Learning human arm movements by imitation: Evaluation of a biologically-inspired connectionist architecture, *in* "Proceedings of the First IEEE-RAS International Conference on Humanoid Robots (Humanoids2000)," Cambridge, MA.

Blair, P. (1949). *Animation: Learning How to Draw Animated Cartoons,* Walter T. Foster Art Books, Laguna Beech, CA.

Blumberg, B. (1994). Action selection in Hamsterdam: Lessons from ethology, *in* "Proceedings of the Third International Conference on the Simulation of Adaptive Behavior (SAB94)," MIT Press, Cambridge, MA, pp. 108–117.

Blumberg, B. (1996). Old Tricks, New Dogs: Ethology and Interactive Creatures, PhD thesis, Massachusetts Institute of Technology, Media Arts and Sciences.

Blumberg, B., Todd, P., and Maes, M. (1996). No bad dogs: Ethological lessons for learning, *in* "Proceedings of the Fourth International Conference on Simulation of Adaptive Behavior (SAB96)," MIT Press, Cambridge, MA, pp. 295–304.

Brazelton, T. (1979). Evidence of communication in neonatal behavior assessment, *in* M. Bullowa, ed., "Before Speech: The Beginning of Interpersonal Communication," Cambridge University Press, Cambridge, UK, pp. 79–88.

Breazeal, C. (1998). A motivational system for regulating human-robot interaction, *in* "Proceedings of the Fifteenth National Conference on Artificial Intelligence (AAAI98)," Madison, WI, pp. 54–61.

Breazeal, C. (2000a). Believability and readability of robot faces, *in* "Proceedings of the Eighth International Symposium on Intelligent Robot Systems (SIRS2000)," Reading, UK, pp. 247–256.

Breazeal, C. (2000b). Proto-conversations with an anthropomorphic robot, *in* "Proceedings of the Ninth International Workshop on Robot and Human Interactive Communication (RO-MAN2000)," Osaka, Japan, pp. 328–333.

Breazeal, C. (2000c). Sociable Machines: Expressive Social Exchange Between Humans and Robots, PhD thesis, Massachusetts Institute of Technology, Department of Electrical Engineering and Computer Science, Cambridge, MA.

Breazeal, C. (2001a). Affective interaction between humans and robots, *in* "Proceedings of the Sixth European Conference on Artificial Life (ECAL2001)," Prague, Czech Republic, pp. 582–591.

Breazeal, C. (2001b). Socially intelligent robots: Research, development, and applications, *in* "IEEE Conference on Systems, Man, and Cybernetics (SMC2001)," Tuscon, AZ.

Breazeal, C. (2002). Designing sociable robots: Issues and lessons, *in* K. Dautenhahn, A. Bond and L. Canamero, eds., "Socially Intelligent Agents: Creating Relationships with Computers and Robots," Kluwer Academic Press.

Breazeal, C., and Aryananda, L. (2002). "Recognition of affective communicative intent in robot-directed speech," *Autonomous Robots* 12(1), 83–104.

Breazeal, C., Edsinger, A., Fitzpatrick, P., Scassellati, B., and Varchavskaia, P. (2000). "Social constraints on animate vision," *IEEE Intelligent Systems* 15(4), 32–37.

Breazeal, C., Fitzpatrick, P., and Scassellati, B. (2001). "Active vision systems for sociable robots," *IEEE Transactions on Systems, Man, and Cybernetics: Special Issue Part A, Systems and Humans.* K. Dautenhahn (ed.).

Breazeal, C., and Foerst, A. (1999). Schmoozing with robots: Exploring the original wireless network, *in* "Proceedings of Cognitive Technology (CT99)," San Francisco, CA, pp. 375–390.

Breazeal, C., and Scassellati, B. (1999a). A context-dependent attention system for a social robot, *in* "Proceedings of the Sixteenth International Joint Conference on Artificial Intelligence (IJCAI99)," Stockholm, Sweden, pp. 1146–1151.

Breazeal, C., and Scassellati, B. (1999b). How to build robots that make friends and influence people, *in* "Proceedings of the 1999 IEEE/RSJ International Conference on Intelligent Robots and Systems (IROS99)," Kyonju, Korea, pp. 858–863.

Breazeal, C., and Scassellati, B. (2000). "Infant-like social interactions between a robot and a human caretaker," *Adaptive Behavior* 8(1), 47–72.

Breazeal, C., and Scassellati, B. (2002). Challenges in building robots that imitate people, *in* K. Dautenhahn and C. Nehaniv, eds., "Imitation in Animals and Artifacts," MIT Press.

Brooks, R. (1986). "A robust layered control system for a mobile robot," *IEEE Journal of Robotics and Automation* RA-2, 253–262.

Brooks, R. (1990). Challenges for complete creature architectures, *in* "Proceedings of the First International Conference on Simulation of Adaptive Behavior (SAB90)," MIT Press, Cambridge MA, pp. 434–443.

Brooks, R., Breazeal, C., Marjanovic, M., Scassellati, B., and Williamson, M. (1999). The Cog Project: Building a humanoid robot, *in* C. L. Nehaniv, ed., "Computation for Metaphors, Analogy and Agents," Vol. 1562 of *Springer Lecture Notes in Artificial Intelligence,* Springer-Verlag, New York, NY.

Brothers, L. (1997). *Friday's Footprint: How Society Shapes the Human Mind,* Oxford University Press, New York, NY.

Bruner, J. (1991). "The Narrative construction of reality," *Critical Inquiry* 18, 1–21.

Bullowa, M. (1979). *Before Speech: The Beginning of Interpersonal Communication,* Cambridge University Press, Cambridge, UK.

Burgard, W., Cremers, A., Fox, D., Haehnel, D., Lakemeyer, G., Schulz, D., Steiner, W., and Thrun, S. (1998). The interactive museum tour-guide robot, *in* "Proceedings of the Fifthteenth National Conference on Artificial Intelligence (AAAI98)," Madison, WI, pp. 11–18.

Cahn, J. (1990). Generating Expression in Synthesized Speech, Master's thesis, Massachusetts Institute of Technology, Media Arts and Science, Cambridge, MA.

Canamero, D. (1997). Modeling motivations and emotions as a basis for intelligent behavior, *in* L. Johnson, ed., "Proceedings of the First International Conference on Autonomous Agents (Agents97)," ACM Press, pp. 148–155.

Carey, S., and Gelman, R. (1991). *The Epigenesis of Mind,* Lawrence Erlbaum Associates, Hillsdale, NJ.

Caron, A., Caron, R., Caldwell, R., and Weiss, S. (1973). "Infant perception of structural properties of the face," *Developmental psychology* 9, 385–399.

Carver, C., and Scheier, M. (1998). *On the Self-Regulation of Behavior,* Cambridge University Press, Cambridge, UK.

Cassell, J. (1999a). Embodied conversation: Integrating face and gesture into automatic spoken dialog systems, *in* Luperfoy, ed., "Spoken Dialog Systems," MIT Press, Cambridge, MA.

Cassell, J. (1999b). Nudge nudge wink wink: Elements of face-to-face conversation for embodied conversational agents, *in* J. Cassell, J. Sullivan, S. Prevost and E. Churchill, eds., "Embodied Conversational Agents," MIT Press, Cambridge, MA, pp. 1–27.

Cassell, J., Bickmore, T., Campbell, L., Vilhjalmsson, H., and Yan, H. (2000). Human conversation as a system framework: Designing embodied conversation agents, *in* J. Cassell, J. Sullivan, S. Prevost and E. Churchill, eds., "Embodied Conversational Agents," MIT Press, Cambridge, MA, pp. 29–63.

Cassell, J., and Thorisson, K. (1999). "The power of a nod and a glance: Envelope versus emotional feedback in animated conversational agents," *Applied Artificial Intelligence* 13, 519–538.

Chen, L., and Huang, T. (1998). Multimodal human emotion/expression recognition, *in* "Proceedings of the Second International Conference on Automatic Face and Gesture Recognition," Nara, Japan, pp. 366–371.

Cole, J. (1998). *About Face,* MIT Press, Cambridge, MA.

Collis, G. (1979). Describing the structure of social interaction in infancy, *in* M. Bullowa, ed., "Before Speech: The Beginning of Interpersonal Communication," Cambridge University Press, Cambridge, UK, pp. 111–130.

Damasio, A. (1994). *Descartes Error: Emotion, Reason, and the Human Brain,* G.P. Putnam's Sons, New York, NY.

Damasio, A. (1999). *The Feeling of What Happens: Body and Emotion in the Making of Consciousness,* Harcourt Brace, New York, NY.

Dario, P., and Susani, G. (1996). Physical and psychological interactions between humans and robots in the home environment, *in* "Proceedings of the First International Symposium on Humanoid Robots (HURO96)," Tokyo, Japan, pp. 5–16.

Darwin, C. (1872). *The Expression of the Emotions in Man and Animals,* John Murray, London, UK.

Dautenhahn, K. (1997). "I could be you—the phenomenological dimension of social understanding," *Cybernetics and Systems Journal* 28(5), 417–453.

Dautenhahn, K. (1998). "The art of designing socially intelligent agents: Science, fiction, and the human in the loop," *Applied Artificial Intelligence Journal* 12(7–8), 573–617.

Dautenhahn, K. (1999a). Embodiment and interaction in socially intelligent life-like agents, *in* C. L. Nehaniv, ed., "Computation for Metaphors, Analogy and Agents," Vol. 1562 of *Springer Lecture Notes in Artificial Intelligence,* Springer-Verlag, pp. 102–142.

Dautenhahn, K. (1999b). The lemur's tale—Story-telling in primates and other socially intelligent agents, *in* M. Mateas and P. Sengers, eds., "AAAI Fall Symposium on Narrative Intelligence," AAAI Press, pp. 59–66.

Dautenhahn, K. (2000). Design issues on interactive environments for children with autism, *in* "Proceedings of the Third International Conference on Disability, Virtual Reality and Associated Technologies (ICDVRAT 2000)," Alghero Sardinia, Italy, pp. 153–161.

Dautenhahn, K., and Coles, S. (2001). "Narrative intelligence from the bottom up: A computational framework for the study of story-telling in autonomous agents," *Journal of Artificial Societies and Social Simulation (JASSS).*

de Boysson-Bardies, B. (1999). *How Language Comes to Children, from Birth to Two Years,* MIT Press, Cambridge, MA.

Dellaert, F., Polzin, F., and Waibel, A. (1996). Recognizing emotion in speech, *in* "Proceedings of the 1996 International Conference on Spoken Language Processing (ICSLP96)."

Dennett, D. (1987). *The Intentional Stance,* MIT Press, Cambridge, MA.

Dick, P. (1990). *Blade Runner (Do Androids Dream of Electric Sheep?),* Ballantine Books, New York, NY.

Dickinson, T., Michael, K., Kauer, J., and Walt, D. (1999). "Convergent, self-encoded bead sensor arrays in the design of an artificial nose," *Analytical Chemistry* pp. 2192–2198.

Duchenne, B. (1990). *The Mechanism of Human Facial Expression,* Cambridge University Press, New York, NY. translated by R. A. Cuthbertson.

Eckerman, C., and Stein, M. (1987). "How imitation begets imitation and toddlers' generation of games," *Developmental Psychology* 26, 370–378.

Eibl-Eibesfeldt, I. (1972). Similarities and differences between cultures in expressive movements, *in* R. Hinde, ed., "Nonverbal Communication," Cambridge University Press, Cambridge, UK, pp. 297–311.

Ekman, P. (1992). "Are there basic emotions?," *Psychological Review* 99(3), 550–553.

Ekman, P., and Friesen, W. (1982). Measuring facial movement with the Facial Action Coding System, *in* "Emotion in the Human Face," Cambridge University Press, Cambridge, UK, pp. 178–211.

Ekman, P., Friesen, W., and Ellsworth, P. (1982). What emotion categories or dimensions can observers judge from facial behavior?, *in* P. Ekman, ed., "Emotion in the Human Face," Cambridge University Press, Cambridge, UK, pp. 39–55.

Ekman, P., and Oster, H. (1982). Review of research, 1970 to 1980, *in* P. Ekman, ed., "Emotion in the Human Face," Cambridge University Press, Cambridge, UK, pp. 147–174.

Elliot, C. D. (1992). The Affective Reasoner: A Process Model of Emotions in a Multi-Agent System, PhD thesis, Northwestern University, Institute for the Learning Sciences, Chicago, IL.

Faigin, G. (1990). *The Artist's Complete Guide to Facial Expression,* Watson Guptill, New York, NY.

Fantz, R. (1963). "Pattern vision in newborn infants," *Science* 140, 296–297.

Fernald, A. (1984). The perceptual and affective salience of mothers' speech to infants, *in* C. G. L. Feagans and R. Golinkoff, eds., "The Origins and Growth of Communication," Ablex Publishing, Norwood, NJ, pp. 5–29.

Fernald, A. (1989). "Intonation and communicative intent in mother's speech to infants: Is the melody the message?," *Child Development* 60, 1497–1510.

Fernald, A. (1993). "Approval and disapproval: Infant responsiveness to vocal affect in familiar and unfamiliar languages," *Developmental Psychology* 64, 657–674.

Ferrier, L. (1985). Intonation in discourse: Talk between 12-month-olds and their mothers, *in* K. Nelson, ed., "Children's Language," Lawrence Erlbaum Associates, Hillsdale, NJ, pp. 35–60.

Fleming, B., and Dobbs, D. (1999). *Animating Facial Features and Expressions,* Charles river media, Rockland, MA.

Foerst, A. (1999). "Artificial sociability: From embodied AI toward new understandings of personhood," *Technology in Society* pp. 373–386.

Foerst, A., and Petersen, L. (1998). Identity, formation, dignity: The impacts of artificial intelligence upon Jewish and Christian understandings of personhood, *in* "Proceedings of AAR98," Orlando, FL.

Forgas, J. (2000). *Affect and Social Cognition,* Lawrence Erlbaum Associates, Hillsdale, NJ.

Frijda, N. (1969). Recognition of emotion, *in* L. Berkowitz, ed., "Advances in Experimental Social Psychology," Academic Press, New York, NY, pp. 167–223.

Frijda, N. (1994a). Emotions are functional, most of the time, *in* P. Ekman and R. Davidson, eds., "The Nature of Emotion," Oxford University Press, New York, NY, pp. 112–122.

Frijda, N. (1994b). Emotions require cognitions, even if simple ones, *in* P. Ekman and R. Davidson, eds., "The Nature of Emotion," Oxford University Press, New York, NY, pp. 197–202.

Frijda, N. (1994c). Universal antecedents exist, and are interesting, *in* P. Ekman and R. Davidson, eds., "The Nature of Emotion," Oxford University Press, New York, NY, pp. 146–149.

Fujita, M., and Kageyama, K. (1997). An open architecture for robot entertainment, *in* "Proceedings of the First International Conference on Autonomous Agents (Agents97)."

Galef, B. (1988). Imitation in animals: History, definitions, and interpretation of data from the psychological laboratory, *in* T. Zentall and G. Galef, eds., "Social Learning: Psychological and Biological Perspectives," Lawrence Erlbaum Associates, Hillsdale, NJ.

Gallistel, C. (1980). *The Organization of Action,* MIT Press, Cambridge, MA.

Gallistel, C. (1990). *The Organization of Learning,* MIT Press, Cambridge, MA.

Garvey, C. (1974). "Some properties of social play," *Merrill-Palmer Quarterly* 20, 163–180.

Glos, J., and Cassell, J. (1997). Rosebud: A place for interaction between memory, story, and self, *in* "Proceedings of the Second International Cognitive Technology Conference (CT99)," IEEE Computer Society Press, San Francisco, pp. 88–97.

Gould, J. (1982). *Ethology,* Norton.

Grieser, D., and Kuhl, P. (1988). "Maternal speech to infants in a tonal language: Support for universal prosodic features in motherese," *Developmental Psychology* 24, 14–20.

Halliday, M. (1975). *Learning How to Mean: Explorations in the Development of Language,* Elsevier, New York, NY.

Halliday, M. (1979). One child's protolanguage, *in* M. Bullowa, ed., "Before Speech: The Beginning of Interpersonal Communication," Cambridge University Press, Cambridge, UK, pp. 149–170.

Hara, F. (1998). Personality characterization of animate face robot through interactive communication with human, *in* "Proceedings of the 1998 International Advanced Robotics Program (IARP98)," Tsukuba, Japan, pp. IV–1.

Hauser, M. (1996). *The Evolution of Communication,* MIT Press, Cambridge, MA.

Hayes, G., and Demiris, J. (1994). A robot controller using learning by imitation, *in* "Proceedings of the Second International Symposium on Intelligent Robotic Systems," Grenoble, France, pp. 198–204.

Hendriks-Jansen, H. (1996). *Catching Ourselves in the Act,* MIT Press, Cambridge, MA.

Hirai, K. (1998). Humanoid robot and its applications, *in* "Proceedings of the 1998 International Advanced Robot Program (IARP98)," Tsukuba, Japan, pp. V–1.

Hirsh-Pasek, K., Jusczyk, P., Cassidy, K. W., Druss, B., and Kennedy, C. (1987). "Clauses are perceptual units of young infants," *Cognition* 26, 269–286.

Horn, B. (1986). *Robot Vision,* MIT Press, Cambridge, MA.

Irie, R. (1995). Robust Sound Localization: An application of an Auditory Perception System for a Humanoid Robot, Master's thesis, Massachusetts Institute of Technology, Department of Electrical Engineering and Computer Science, Cambridge, MA.

Itti, L., Koch, C., and Niebur, E. (1998). "A Model of saliency-based visual attention for rapid scene analysis," *IEEE Transactions on Pattern Analysis and Machine Intelligence (PAMI)* 20(11), 1254–1259.

Izard, C. (1977). *Human Emotions,* Plenum Press, New York, NY.

Izard, C. (1993). "Four systems for emotion activation: Cognitive and noncognitive processes," *Psychological Review* 100, 68–90.

Izard, C. (1994). Cognition is one of four types of emotion activating systems, *in* P. Ekman and R. Davidson, eds., "The Nature of Emotion," Oxford University Press, New York, NY, pp. 203–208.

Johnson, M. (1993). Constraints on cortical plasticity, *in* "Brain Development and Cognition: A Reader," Blackwell, Oxford, UK, pp. 703–721.

Johnson, M., Wilson, A., Blumberg, B., Kline, C., and Bobick, A. (1999). Sympathetic interfaces: Using a plush toy to direct synthetic characters, *in* "Proceedings of the 1999 Conference on Computer Human Interaction (CHI99)," The Hague, The Netherlands.

Kandel, E., Schwartz, J., and Jessell, T. (2000). *Principles of Neuroscience,* third ed., Appelton and Lange, Norwalk, CT.

Kawamura, K., Wilkes, D., Pack, T., Bishay, M., and Barile, J. (1996). Humanoids: future robots for home and factory, *in* "Proceedings of the First International Symposium on Humanoid Robots (HURO96)," Tokyo, Japan, pp. 53–62.

Kaye, K. (1979). Thickening thin data: The maternal role in developing communication and language, *in* M. Bullowa, ed., "Before Speech: The Beginning of Interpersonal Communication," Cambridge University Press, Cambridge, UK, pp. 191–206.

Kinny, D., Georgeff, M., and Rao, A. (1996). A methodology and modelling technique for systems of BDI agents, *in* W. V. de Velde and J. Perram, eds., "Agents Breaking Away: Proceedings of the Seventh European Workshop on Modelling Autonomous Agents in a Multi-Agent World," Eindhoven, The Netherlands, pp. 56–71.

Kitano, H., Tambe, M., Stone, P., Veloso, M., Coradeschi, S., Osawa, E., Matsubara, H., Noda, I., and Asada, M. (1997). The Robocup'97 synthetic agents challenge, *in* "Proceedings of the The First International Workshop on RoboCup as part of IJCAI-97," Nagoya, Japan.

Kolb, B., Wilson, B., and Laughlin, T. (1992). "Developmental changes in the recognition and comprehension of facial expression: Implications for frontal lobe function," *Brain and Cognition* pp. 74–84.

Kuniyoshi, Y., Kita, N., Sugimoto, K., Nakamura, S., and Suehiro, T. (1995). A foveated wide angle lens for active vision, *in* "Proceedings of IEEE International Conference on Robotics and Automation."

Kurzweil, R. (2000). *The Age of Spiritual Machines,* Penguin, New York, NY.

Lakoff, G. (1990). *Women, Fire, and Dangerous Things: What Categories Reveal about the Mind,* University Chicago Press, Chicago, IL.

Lazarus, R. (1991). *Emotion and Adaptation,* Oxford University Press, New York, NY.

Lazarus, R. (1994). Universal antecedents of the emotions, *in* P. Ekman and R. Davidson, eds., "The Nature of Emotion," Oxford University Press, New York, NY, pp. 163–171.

Leslie, A. (1994). ToMM, ToBY, and agency: Core architecture and domain specificity, *in* L. Hirschfeld and S. Gelman, eds., "Mapping the Mind: Domain Specificity in cognition and Culture," Cambridge University Press, Cambridge, UK, pp. 119–148.

Lester, J., Towns, S., Callaway, S., Voerman, J., and Fitzgerald, P. (2000). Deictic and emotive communication in animated pedagogical agents, *in* J. Cassell, J. Sullivan, S. Prevost and E. Churchill, eds., "Embodied Conversational Agents," MIT Press, Cambridge, MA, pp. 123–154.

Levenson, R. (1994). Human emotions: A functional view, *in* P. Ekman and R. Davidson, eds., "The Nature of Emotion," Oxford University Press, New York, NY, pp. 123–126.

Lorenz, K. (1973). *Foundations of Ethology,* Springer-Verlag, New York, NY.

Madsen, R. (1969). *Animated Film: Concepts, Methods, Uses,* Interland, New York.

Maes, P. (1991). Learning behavior networks from experience, *in* "Proceedings of the First European Conference on Artificial Life (ECAL90)," MIT Press, Paris, France.

Maes, P., Darrell, T., Blumberg, B., and Pentland, A. (1996). The ALIVE system: wireless, full-body interaction with autonomous agents, *in* "ACM Multimedia Systems," ACM Press, New York, NY.

Maratos, O. (1973). The Origin and Development of Imitation in the First Six Months of Life, PhD thesis, University of Geneva.

Mataric, M., Williamson, M., Demiris, J., and Mohan, A. (1998). Behavior-based primitives for articulated control, *in* "Proceedings of the Sixth International Conference on Simulation of Adaptive Behavior (SAB98)," Zurich, Switzerland, pp. 165–170.

Matsusaka, Y., and Kobayashi, T. (1999). Human interface of humanoid robot realizing group communication in real space, *in* "Proceedings of the Second International Symposium on Humanoid Robots (HURO99)," Tokyo, Japan, pp. 188–193.

McFarland, D., and Bosser, T. (1993). *Intelligent Behavior in Animals and Robots,* MIT Press, Cambridge, MA.

McRoberts, G., Fernald, A., and Moses, L. (2000). "An acoustic study of prosodic form-function relationships infant-directed speech," *Developmental Psychology.*

Mead, G. (1934). *Mind, Self, and Society,* Univeristy of Chicago Press.

Meltzoff, A., and Moore, M. (1977). "Imitation of facial and manual gestures by human neonates," *Science* 198, 75–78.

Mills, M., and Melhuish, E. (1974). "Recognition of mother's voice in early infancy," *Nature* 252, 123–124.

Minsky, M. (1988). *The Society of Mind,* Simon and Schuster, New York.

Mithen, S. (1996). *The Prehistory of the Mind,* Thames and Hudson Ltd., London.

Mizuuchi, I., Hara, A., Inaba, M., and Inoue, H. (2000). Tendon-driven torso control for a whole-body agent which has multi-DOF spine, *in* "Proceedings of the Eighteenth Annual Conference of the Robotics Society of Japan," Vol. 3, pp. 1459–1460.

Mumme, D., Fernald, A., and Herrera, C. (1996). "Infants' response to facial and vocal emotional signals in a social referencing paradigm," *Child Development* 67, 3219–3237.

Murray, I., and Arnott, L. (1993). "Toward the simulation of emotion in synthetic speech: A review of the literature on human vocal emotion," *Journal Acoustical Society of America* 93(2), 1097–1108.

Nakatsu, R., Nicholson, J., and Tosa, N. (1999). Emotion recognition and its application to computer agents with spontaneous interactive capabilities, *in* "Proceedings of the 1999 International Conference on Multimedia Computing and Systems (ICMCS99)," Vol. 2, Florence, Italy, pp. 804–808.

Nelson, K. (1993). "The psychological and social origins of autobiographical memory," *Psychological Science* 4(1), 7–14.

Newman, R., and Zelinsky, A. (1998). Error analysis of head pose and gaze direction from stereo vision, *in* "Proceedings of 1998 IEEE/RSJ International Conference on Intelligent Robots and Systems (IROS98)," Victoria B.C., Canada, pp. 527–532.

Newson, J. (1979). The growth of shared understandings between infant and caregiver, *in* M. Bullowa, ed., "Before Speech: The Beginning of Interpersonal Communication," Cambridge University Press, Cambridge, UK, pp. 207–222.

Niedenthal, P., and Kityama, S. (1994). *The Heart's Eye: Emotional Influences in Perception and Attention,* Academic Press, San Diego.

Nothdurft, H. C. (1993). "The role of features in preattentive vision: Comparison of orientation, motion and color cues," *Vision Research* 33, 1937–1958.

Okada, M., Nakamura, Y., and Ban, S. (2000). Design of Programmable Passive Compliance Mechanism using Closed Kinematic Chain—PPC cybernetic shoulder for humanoid robots, *in* "Proceedings of the 2000 International Symposium on Experimental Robotics (ISER00)," Honolulu, HI, pp. 31–40.

Ortony, A., Clore, G., and Collins, A. (1988) *The Cognitive Structure of Emotion,* Cambridge University Press, Cambridge, UK.

Otake, M., Inaba, M., and Inoue, H. (1999). Development of gel robots made of electro-active polymer PAMPS gel, *in* "Proceedings of the 1999 International Conference on Systems, Man and Cybernetics (IEEE SMC99)," Tokyo, Japan, pp. 788–793.

Papousek, M., Papousek, H., and Bornstein, M. (1985). The naturalistic vocal environment of young infants: On the significance of homogeneity and variability in parental speech, *in* T. Field and N. Fox, eds., "Social Perception in Infants," Ablex, Norwood, NJ, pp. 269–297.

Parke, F. (1972). Computer Generated Animation of Faces, PhD thesis, University of Utah, Salt Lake City. UTEC-CSc-72-120.

Parke, F., and Waters, K. (1996). *Computer Facial Animation,* A. K. Peters, Wellesley, MA.

Pepperberg, I. (1988). "An interactive modeling technique for acquisition of communication skills: Separation of "labeling" and "requesting" in a psittachine subject," *Applied Psycholinguistics* 9, 59–76.

Pepperberg, I. (1990). "Referential mapping: A technique for attaching functional significance to the innovative utterances of an african grey parrot," *Applied Psycholinguistics* 11, 23–44.

Picard, R. (1997). *Affective Computation,* MIT Press, Cambridge, MA.

Plutchik, R. (1984). Emotions: A general psychevolutionary theory, *in* K. Scherer and P. Elkman, eds., "Approaches to Emotion," Lawrence Erlbaum Associates, Hillsdale, New Jersey, pp. 197–219.

Plutchik, R. (1991). *The Emotions,* University Press of America, Lanham, MD.

Pratt, G., and Williamson, M. (1995). Series elastic actuators, *in* "Proceedings of the 1995 International Conference on Intelligent Robots and Systems (IROS95)," Pittsberg, PA.

Premack, D., and Premack, A. (1995). Origins of human social competence, *in* M. Gazzaniga, ed., "The Cognitive Neurosciences," Bradford, New York, NY, pp. 205–218.

Redican, W. (1982). An evolutionary perspective on human facial displays, *in* "Emotion in the Human Face," Cambridge University Press, Cambridge, UK, pp. 212–280.

Reeves, B., and Nass, C. (1996). *The Media Equation,* CSLI Publications, Stanford, CA.

Reilly, S. (1996). Believable Social and Emotional Agents, PhD thesis, Carnegie Mellon University, School of Computer Science, Pittsburgh, PA.

Rhodes, B. (1997). "The wearable remembrance agent: A system for augmented memory," *Personal Technologies.*

Rickel, J., and Johnson, W. L. (2000). Task-oriented collaboration with embodied agents in virtual worlds, *in* J. Cassell, J. Sullivan, S. Prevost and E. Churchill, eds., "Embodied Conversational Agents," MIT Press, Cambridge, MA, pp. 95–122.

Robinson, D., Pratt, J., Paluska, D., and Pratt, G. (1999). Series elastic actuator development for a biomimetic walking Robot, *in* "Proceedings of IEEE/ASME International Conference on Advanced Intelligent Mechatronics," Atlanta, GA.

Roy, D., and Pentland, A. (1996). Automatic spoken affect analysis and classification, *in* "Proceedings of the 1996 International Conference on Automatic Face and Gesture Recognition."

Russell, J. (1997). Reading emotions from and into faces: Resurrecting a dimensional-contextual perspective, *in* J. Russell and J. Fernandez-Dols, eds., "The psychology of Facial Expression," Cambridge university press, Cambridge, UK, pp. 295–320.

Rutter, D., and Durkin, K. (1987). "Turn-taking in mother-infant interaction: An examination of volications and gaze," *Developmental Psychology* 23(1), 54–61.

Sanders, G., and Scholtz, J. (2000). Measurement and evaluation of embodied conversational agents, *in* J. Cassell, J. Sullivan, S. Prevost and E. Churchill, eds., "Embodied Conversational Agents," MIT Press, Cambridge, MA, pp. 346–373.

Scassellati, B. (1998). Finding eyes and faces with a foveated vision system, *in* "Proceedings of the Fifthteenth National Conference on Artificial Intelligence (AAAI98)," Madison, WI, pp. 969–976.

Scassellati, B. (1999). Imitation and mechanisms of joint attention: A developmental structure for building social skills on a humanoid robot, *in* C. L. Nehaniv, ed., "Computation for Metaphors, Analogy and Agents," Vol. 1562 of *Springer Lecture Notes in Artificial Intelligence,* Springer-Verlag, New York, NY.

Scassellati, B. (2000a). A theory of mind for a humanoid robot, *in* "Proceedings of the First IEEE-RAS International Conference on Humanoid Robots (Humanoids2000)," Cambridge, MA.

Scassellati, B. (2000b). Theory of mind...for a robot, *in* "Proceedings of the 2000 AAAI Fall Symposium on Socially Intelligent Agents—The Human in the Loop," Cape Cod, MA, pp. 164–167. Technical Report FS-00-04.

Schaal, S. (1997). Learning from demonstration, *in* "Proceedings of the 1997 Conference on Neural Information Processing Systems (NIPS97)," Denver, CO, pp. 1040–1046.

Schaal, S. (1999). "Is imitation learning the route to humanoid robots?," *Trends in Cognitive Science* 3(6), 233–242.

Schaffer, H. (1977). Early interactive development, *in* "Studies of Mother-Infant Interaction: Proceedings of Loch Lomonds Symposium," Academic Press, New York, NY, pp. 3–18.

Schank, R., and Abelson, R. (1977). *Scripts, Plans, Goals and Understanding: An Inquiry into Human Knowledge Structure,* Lawrence Erlbaum Associates, Hillsdale, NJ.

Scherer, K. (1984). On the nature and function of emotion: A component process approach, *in* K. Scherer, and P. Ekman, eds, "Approaches to Emotion," Lawrence Erlbaum Associates, Hillsdale, NJ, pp. 293–317.

Scherer, K. (1994). Evidence for both universality and cultural specificity of emotion elicitation, *in* P. Ekman and R. Davidson, eds., "The Nature of Emotion," Oxford University Press, New York, NY, pp. 172–175.

Siegel, D. (1999). *The Developing Mind: Toward a Neurobiology of Interpersonal Experience,* The Guilford Press, New York, NY.

Sinha, P. (1994). "Object recognition via image invariants: A case study," *Investigative Ophthalmology and Visual Science* 35, 1735–1740.

Slaney, M., and McRoberts, G. (1998). "Baby ears: A recognition system for affective vocalizations," *Proceedings of the 1998 International Conference on Acoustics, Speech, and Signal Processing.*

Smith, C. (1989). "Dimensions of appraisal and physiological response in emotion," *Journal of Personality and Social Psychology* 56, 339–353.

Smith, C., and Scott, H. (1997). A componential approach to the meaning of facial expressions, *in* J. Russell and J. Fernandez-Dols, eds., "The Psychology of Facial Expression," Cambridge University Press, Cambridge, UK, pp. 229–254.

Snow, C. (1972). "Mother's speech to children learning language," *Child Development* 43, 549–565.

Stephenson, N. (2000). *The Diamond Age,* Bantam Doubleday Dell Publishers, New York, NY.

Stern, D. (1975). Infant regulation of maternal play behavior and/or maternal regulation of infant play behavior, *in* "Proceedings of the Society of Research in Child Development."

Stern, D., Spieker, S., and MacKain, K. (1982). "Intonation contours as signals in maternal speech to prelinguistic infants," *Developmental Psychology* 18, 727–735.

Takanobu, H., Takanishi, A., Hirano, S., Kato, I., Sato, K., and Umetsu, T. (1999). Development of humanoid robot heads for natural human-robot communication, *in* "Proceedings of Second International Symposium on Humanoid Robots (HURO99)," Tokyo, Japan, pp. 21–28.

Takeuchi, A., and Nagao, K. (1993). Communicative facial displays as a new conversational modality, *in* "Proceedings of the 1993 ACM Conference on Human Factors in Computing Systems (ACM SIGCHI93)," Amsterdam, The Netherlands, pp. 187–193.

Thomas, F., and Johnston, O. (1981). *Disney Animation: The Illusion of Life,* Abbeville Press, New York.

Thorisson, K. (1998). Real-time decision making in multimodal face-to-face communication, *in* "Second International Conference on Autonomous Agents (Agents98)," Minneapolis, MN, pp. 16–23.

Thrun, S., Bennewitz, M., Burgard, W., Cremers, A., Dellaert, F., Fox, D., Haehnel, D., Rosenberg, C., Roy, N., Schulte, J., and Schulz, D. (1999). MINERVA: A second generation mobile tour-guide robot, *in* "Proceedings of the IEEE International Conference on Robotics and Automation (ICRA'99)," Detroit, MI.

Tinbergen, N. (1951). *The Study of Instinct,* Oxford University Press, New York, NY.

Trehub, S., and Trainor, L. (1990). *Rules for Listening in Infancy,* Elsevier, North Holland, chapter 5.

Trevarthen, C. (1979). Communication and cooperation in early infancy: A description of primary intersubjectivity, *in* M. Bullowa, ed., "Before Speech: The Beginning of Interpersonal Communication," Cambridge University Press, Cambridge, UK, pp. 321–348.

Triesman, A. (1986). "Features and objects in visual processing," *Scientific American* 225, 114B–125.

Tronick, E., Als, H., and Adamson, L. (1979). Structure of early face-to-face communicative interactions, *in* M. Bullowa, ed., "Before Speech: The Beginning of Interpersonal Communication," Cambridge University Press, Cambridge, UK, pp. 349–370.

Tyrrell, T. (1994). "An evaluation of Maes's bottom-up mechanism for behavior selection," *Adaptive Behavior* 2(4), 307–348.

Ude, A., Man, C., Riley, M., and Atkeson, C. G. (2000). Automatic generation of kinematic models for the conversion of human motion capture data into humanoid robot motion, *in* "Proceedings of the First IEEE-RAS International Conference on Humanoid Robots (Humanoids2000)," Cambridge, MA.

van der Spiegel, J., Kreider, G., Claeys, C., Debusschere, I., Sandini, G., Dario, P., Fantini, F., Belluti, P., and Soncini, G. (1989). A foveated retina-like sensor using CCD technology, *in* C. Mead and M. Ismail, eds., "Analog VLSI Implementation of Neural Systems," Kluwer Academic Publishers, pp. 189–212.

Velasquez, J. (1998). When robots weep: A mechanism for emotional memories, *in* "Proceedings of the 1998 National Conference on Artificial Intelligence, AAAI98," pp. 70–75.

Veloso, M., Stone, P., Han, K., and Achim, S. (1997). CMUnited: A team of robotic soccer agents collaborating in an adversarial environment, *in* "In Proceedings of the The First International Workshop on RoboCup in IJCAI-97," Nagoya, Japan.

Vilhjalmsson, H., and Cassell, J. (1998). BodyChat: Autonomous communicative behaviors in avatars, *in* "Proceedings of the Second Annual Conference on Automous Agents (Agents98)," Minneapolis, MN, pp. 269–276.

Vlassis, N., and Likas, A. (1999). "A Kurtosis-based dynamic approach to gaussian mixture modeling," *IEEE Transactions on Systems, Man, and Cybernetics: Part A.*

Vygotsky, L., Vygotsky, S., and John-Steiner, V. (1980). *Mind in Society: The Development of Higher Psychological Processes,* Harvard University Press, Cambridge, MA.

Waters, K., and Levergood, T. (1993). DECface: An automatic lip synchronization algorithm for synthetic faces, Technical Report CRL 94/4, DEC Cambridge Research Laboratory, Cambridge, MA.

Wolfe, J. M. (1994). "Guided Search 2.0: A revised model of visual search," *Psychonomic Bulletin & Review* 1(2), 202–238.

Wood, D., Bruner, J. S., and Ross, G. (1976). "The role of tutoring in problem-solving," *Journal of Child Psychology and Psychiatry* 17, 89–100.

Woodworth, R. (1938). *Experimental Psychology,* Holt, New York.

Yoon, S., Blumberg, B., and Schneider, G. (2000). Motivation driven learning for interactive synthetic characters, *in* "Proceedings of the Fourth International Conference on Autonomous Agents (Agents00)," Barcelona, Spain.

Index